**Advanced Topics in Science and Technology in China**

Volume 2

Zhejiang University is one of the leading universities in China. In Advanced Topics in Science and Technology in China, Zhejiang University Press and Springer jointly publish monographs by Chinese scholars and professors, as well as invited authors and editors from abroad who are outstanding experts and scholars in their fields. This series will be of interest to researchers, lecturers, and graduate students alike.

Advanced Topics in Science and Technology in China aims to present the latest and most cutting-edge theories, techniques, and methodologies in various research areas in China. It covers all disciplines in the fields of natural science and technology, including but not limited to, computer science, materials science, the life sciences, engineering, environmental sciences, mathematics, and physics.

This book series is indexed by both the **Scopus** and **Compendex** databases.

If you are interested in publishing your book in the series, please contact Violetta Xu (Email: violetta_xuqian@zju.edu.cn) and Mengchu Huang (Email: mengchu.huang@cn.springer.com).

Sichao Du · Wen-Yan Yin

# Photodetection and Image Sensing Techniques

ZHEJIANG UNIVERSITY PRESS
浙江大学出版社

Springer

Sichao Du
Zhejiang University City College
Hangzhou, Zhejiang, China

Wen-Yan Yin
Zhejiang University
Hangzhou, Zhejiang, China

ISSN 1995-6819 ISSN 1995-6827 (electronic)
Advanced Topics in Science and Technology in China
ISBN 978-981-96-5478-9 ISBN 978-981-96-5479-6 (eBook)
https://doi.org/10.1007/978-981-96-5479-6

Jointly published with Zhejiang University Press
The print edition is not for sale in China (Mainland). Customers from China (Mainland) please order the print book from: Zhejiang University Press.



This Springer imprint is published by the registered company Springer Nature Singapore Pte Ltd.
The registered company address is: 152 Beach Road, #21-01/04 Gateway East, Singapore 189721, Singapore

# Preface

In today's rapidly advancing technological landscape, optoelectronic imaging technology has emerged as a vital driving force behind the development of modern science and technology. With its unique appeal and extensive applications, optoelectronic imaging seamlessly integrates optics, electronics, information science, and other multidisciplinary knowledge. This comprehensive technology has not only transformed the way we perceive and understand the world but has also demonstrated significant potential in various fields, including remote sensing, biomedicine, mechanical testing, and safety inspection.

This book aims to provide readers with a thorough and systematic introduction to optoelectronic imaging techniques. Beginning with the fundamentals of photoelectric detection technology, it delves into essential topics such as radiation measurement, light metrics, semiconductor principles, and the physical effects and characteristic parameters of photoelectric detection devices. The book also explores the types and characteristics of light sources commonly used in photoelectric detection, forming a strong foundation for understanding advanced concepts such as single-photon detectors, single-photon imaging technology, and spectral imaging technology.

Regarding single-photon detectors and single-photon imaging technology, the book provides an in-depth analysis of the working principles, classification, and performance evaluation methods of single-photon detectors. It highlights their crucial role in imaging through practical applications. Additionally, the book examines the fundamental principles, system construction, and performance characteristics of single-photon imaging, alongside its cutting-edge applications in low-light detection and quantum communication.

Spectral imaging technology and 2D imaging technology are also explored as two critical branches of the optoelectronic imaging field. The book introduces the basic concepts, system classification, principles, and data processing methods of spectral imaging, enabling readers to grasp its immense application value in areas such as remote sensing and biomedicine. Furthermore, it explains the core principles, system classifications, characteristics, resolution, and dynamic range of 2D imaging, equipping readers with a deeper understanding of its essential elements and application domains.

Beyond the theoretical framework, the book emphasizes the practical applications of photoelectric detection technology across diverse fields. These include remote sensing, AAS analysis, and mechanical quantity detection. Through specific case studies, the book showcases the broad utility and significant potential of photoelectric detection technology. It also delves into the frontiers and emerging trends in opto-electronic imaging technology, shedding light on its future development directions and challenges for readers.

Hangzhou, China

Sichao Du
Wen-Yan Yin

# Contents

# Chapter 1
# Basis of Photoelectric Detection Technology

**Abstract** This chapter discusses key concepts and applications in radiation metrics, photometry, and semiconductor physics, with a focus on their impact on photoelectric detection technologies. The challenges faced by China's semiconductor industry, such as equipment shortages, low yields, and underdeveloped domestic technology, are addressed, with recommendations for strategic investments and international cooperation. The chapter also analyzes mechanisms like the photovoltaic and photothermoelectric effects, and key performance parameters such as sensitivity, response time, and noise equivalent power. It concludes by highlighting research trends in photoelectric materials, pointing toward future developments in high-sensitivity, fast-response, and reliable optoelectronic devices.

**Keywords** Silicon · Doping · Penta-valent · Tri-valent · Photoelectric effect · Valence band · Conduction band · Fermi level · Diode · Transistor · Responsivity

## 1.1 Basics of Radiation and Related Metrics

1. **Radiation measurement**

The radiation metric is a set of physical quantities for quantification of electromagnetic radiation. In the photoelectric detection technology, the following basic radiation metrics are mainly involved:

(1) **Radiation intensity (I)**

The radiation flux emitted by a point radiation source per unit solid angle in a specific direction is denoted as $I_e$, defined as $I_e = d\phi/d\Omega$ where $d\phi$ represents the radiation flux within the solid angle element $d\Omega$. . The SI unit of radiation intensity is watt per steradian (W/sr).

For most radiation sources, the radiation intensity is proportional to the cosine of the angle between the given direction and the surface normal. Such a source is referred to as a cosine emitter. Additionally, the energy of radiation at a specific wavelength,



S. Du and W. Yin, *Photodetection and Image Sensing Techniques*, Advanced Topics in Science and Technology in China 2, https://doi.org/10.1007/978-981-96-5479-6_1

per unit surface area, and per unit time, is also referred to as the monochromatic radiation intensity.

(2) **Radiation flux (Φ)**

Radiation flux, also known as radiation power, refers to the radiation energy passing through a certain section per unit of time. It is the power transmitted, or received in the form of radiation, measured in W (watt), that is, 1 W = J/s (joule per second).

2. **Luminosity**

(1) **Illuminance (L)**

Illuminance is a physical measure of the amount of light energy incident on the surface of an object. It is measured in lux (lx). The concept of illuminance is used to describe the brightness of light on a surface, defined as the distribution of luminous flux per unit area perpendicular to the direction of light propagation.

In practical applications, measuring illuminance is crucial in fields such as lighting design, photography, solar power generation, and more. For instance, in lighting engineering, measuring illuminance ensures adequate brightness to meet the specific lighting requirements of various environments and occasions.

(2) **Light intensity (I)**

Light intensity refers to the luminous flux emitted by a light source within a solid angle in a specific direction. It is measured in candelas (cd). Light intensity is a vector quantity that depends not only on the brightness of the light source but also on the luminous area and the distance from the light source. The concept of light intensity is widely applied in lighting technology, optical design, and laser technology. For example, in lighting design, measuring light intensity allows for evaluating the performance of a light source in different directions, enabling the creation of more efficient and well-designed lighting schemes.

(3) **Optical flux**

Luminous flux refers to the total amount of light energy emitted by a light source in all directions and is measured in lumens (lm). It provides a measure of the overall brightness of a light source and is calculated as the product of the light intensity, the illuminated area, and the solid angle. In practical applications, measuring luminous flux is essential for evaluating the overall performance of a light source. For instance, in lighting engineering, luminous flux measurements help determine the brightness of a light source, enabling the design of effective lighting solutions. Additionally, luminous flux is a key parameter for assessing the efficiency of lamps, typically expressed as the ratio of luminous flux output to the input electrical energy.

3. **Radiation metrics and photometry**

Radiation metrics and photometry are two important frameworks for describing different aspects of light. While they differ in definition, measurement, and application, they are closely related. Radiation metrics deal with the measurement of electromagnetic radiation energy. They focus on the physical properties of electromagnetic radiation, such as radiation intensity, radiation flux, and radiation energy flow density. These are purely physical quantities used to quantify radiation levels and doses. Radiation metrics are widely applied in fields such as nuclear energy, medicine, industry, and environmental science to assess radiation exposure risks and the environmental impact of radiation. Photometry, on the other hand, focuses on measuring the visual stimuli caused by light radiation as perceived by the average human eye. Photometric quantities include luminous flux, luminous intensity, illuminance, and brightness. These are psychophysical quantities, derived from the evaluation of physical quantities based on human visual perception. Photometry plays a significant role in lighting design, optical instruments, visual perception studies, and related fields.

The relationship between radiation metrics and photometry is primarily reflected in the following aspects:

(1) **Radiation metric is the basis of light metric**

Radiation metrics is the science of measuring electromagnetic radiation energy, focusing on the physical properties of electromagnetic radiation such as radiation intensity, radiation flux, and radiation energy flow density. These radiometric parameters provide fundamental information about the physical characteristics of a light source. Photometry builds upon radiation metrics by incorporating the visual sensitivity of the human eye. Photometric quantities, such as luminous flux, luminous intensity, illuminance, and brightness, are derived by multiplying radiometric quantities by the photopic efficiency function (representing the human eye's sensitivity to light at different wavelengths).

Thus, radiation metrics serve as the foundation for photometry, while photometry extends radiation metrics by accounting for the visual perception characteristics of the human eye.

(2) **The application scenarios of radiation and photo metrics are different**

Radiation metrics primarily focus on the energy and intensity of electromagnetic radiation and are widely used in fields such as nuclear energy, medicine, industry, and environmental science. In these areas, radiation metrics help assess the risks posed by radiation sources, monitor radiation levels, and protect both humans and the environment from radiation hazards.

Photometry, on the other hand, is mainly applied in lighting design, optical instruments, visual perception studies, and similar fields. Photometric measurements help evaluate parameters such as brightness and luminous intensity to ensure optimal lighting effects and visual comfort. Additionally, photometric principles are extensively used in photography, film, television, and related industries to assess optical quality and the visual appeal of images.

While radiometric and photometric measurements play different roles in optical science, they are interconnected. Radiation metrics provide the foundation for photometry, which incorporates the visual perception characteristics of the human eye. In practical applications, selecting the appropriate parameters for measurement and evaluation depends on the specific requirements and scenarios.

4. **Application of photoelectric detection technology**

In photoelectric detection technology, radiation metrics and photometric quantities serve as fundamental physical parameters. Accurate measurement and analysis of these quantities enable the detection, transmission, processing, and control of optical signals. For instance, in solar cells, illuminance is a critical parameter. By measuring illuminance, the energy conversion efficiency of solar cells under varying lighting conditions can be evaluated, leading to the optimization of their design and performance. Similarly, in photo detectors, radiation intensity is a key parameter. By measuring radiation intensity, the sensitivity and response of a photo detector to optical signals can be assessed. This is essential for improving photo detector performance, reducing false alarm rates, and increasing detection range. Moreover, radiation metrics and photometric quantities are vital in other photoelectric detection technologies, such as spectral analysis, laser ranging, and optical imaging. These technologies rely on precise measurement and analysis of optical signals to detect, identify, and track target objects.

In summary, radiation metrics and photometric quantities are indispensable in photoelectric detection technology. Accurate measurement and analysis of these parameters support the detection, transmission, and control of optical signals, driving advancements in the development and application of photoelectric technologies.

## 1.2 Basic Knowledge of Semiconductors

1. **Semiconductor structure**

(1) **Sharing of electrons**

A large number of atoms in the crystal come together (e.g. Si: 0.235 nm, $5 \times 10^{22}/\text{cm}^2$), overlapping the shell farthest from the nucleus, a phenomenon called the sharing of electrons, as shown in Fig. 1.1.

(2) **Formation of energy bands**

The sharing of electrons in a solid leads to small energy differences between electrons that were originally in the same energy state. For example, in a solid composed of $N$ atoms, the electrons associated with these atoms initially possess identical energy levels. However, due to the collective interaction of the electrons with their respective nuclei and the surrounding nuclei, their energies become slightly different. As a result, a single electronic energy level splits into multiple closely spaced energy levels under the influence of the $N$ nuclei. These newly formed levels exhibit very small differences in energy, forming a continuous range of energy levels with a certain width, referred to as energy bands. This concept is illustrated in Fig. 1.2.

The energy bands in the atoms can be divided into the following types:

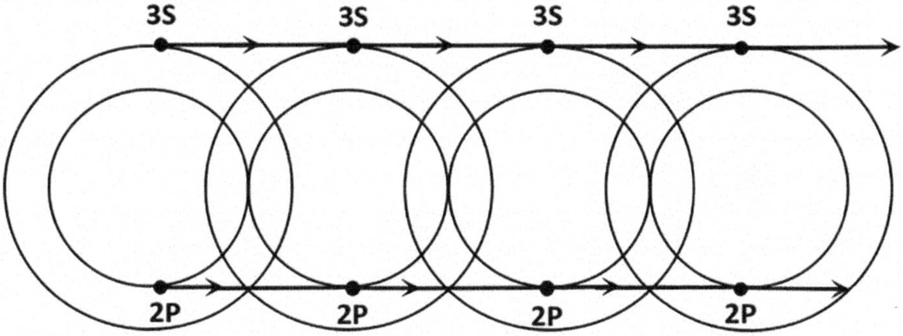


**Fig. 1.1** Schematic representation of the communication motion of the electrons

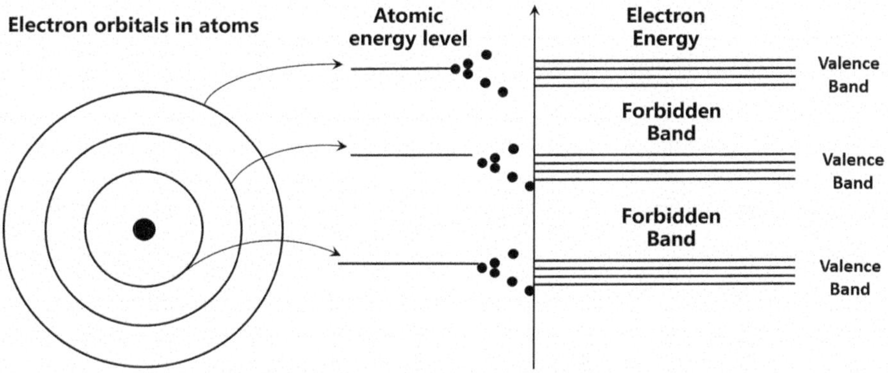


**Fig. 1.2** Schematic diagram of atomic energy-level splitting into energy bands

A. **Forbidden Band**. The energy bands that electrons are allowed to occupy are called allowed bands. The regions between these allowed bands, where electrons are not permitted to reside, are referred to as the forbidden bands.
B. **Valence Band**. The outermost electron in an atom is called the valence electron, and the energy band corresponding to the valence electron level is called the valence band.
C. **Conduction Band**. The conduction band is the lowest energy allowed band that lies above the valence band.

(3) **The energy band structure of semiconductors**

In an intrinsic semiconductor, the electrons in covalent bonds are weakly bound and can gain enough energy to jump across the forbidden band into the conduction band. When electrons transition from the valence band to the conduction band, they become free electrons, and the vacancies left behind in the valence band become free holes. These free electrons and holes are collectively known as charge carriers.

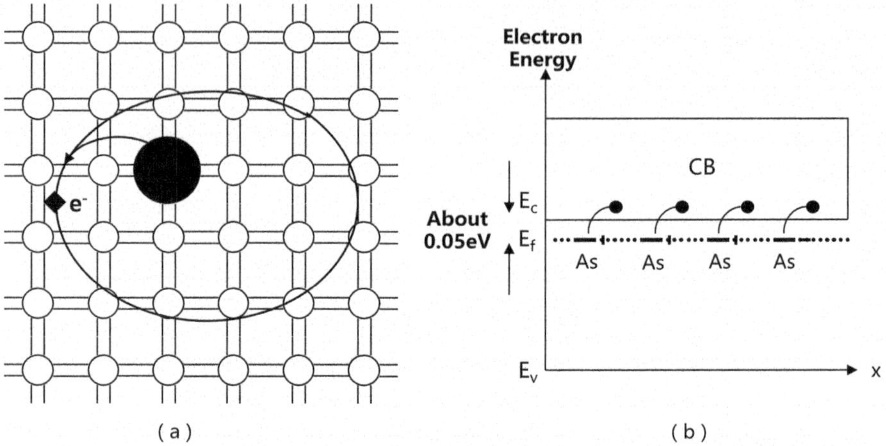


**Fig. 1.3** Structure and band structure of a n-type semiconductor [1]

When trace amounts of penta-valent elements, such as phosphorus, are added to the intrinsic semiconductor, they replace a small number of Si atoms in the crystal lattice, as shown in Fig. 1.3. As depicted in the figure, the outermost phosphorus atom has five valence electrons. Four of these electrons form covalent bonds with nearby Si atoms, while the fifth electron, which is not involved in the covalent bond, is only weakly bound to the phosphorus atom. At room temperature, this electron can gain sufficient energy to become a free electron within the lattice. The phosphorus atoms, after losing their electrons, become immobile positive ions.

Phosphorus atoms are called donor atoms or donor impurities because they can release an electron. In the intrinsic semiconductor, each phosphorus atom generates one free electron, while the number of holes created by intrinsic excitation remains unchanged. As a result, in a phosphorus-doped semiconductor, the number of free electrons greatly exceeds the number of holes, making electrons the majority carriers (also known as negatively charged carriers, or n-type carriers) and holes the minority carriers.

In an intrinsic semiconductor like Si (or germanium(Ge)), when trace amounts of tri-valent elements, such as boron(B), are added, the B atoms replace a small number of Si atoms in the crystal lattice, as shown in Fig. 1.4. As illustrated in the figure, the three valence electrons of a B atom form covalent bonds with three neighboring Si atoms. However, the covalent bond with the fourth neighboring Si atom is incomplete due to the lack of an electron, creating a hole. This hole is filled by a valence electron from a nearby Si atom, while the B atom gains an electron and becomes a negatively charged ion. Consequently, a hole appears in the adjacent covalent bond.

Since B atoms accept electrons, they are called acceptor atoms or acceptor impurities. In the intrinsic semiconductor, each B atom introduces one hole. When a sufficient number of B atoms are incorporated into the semiconductor, the number

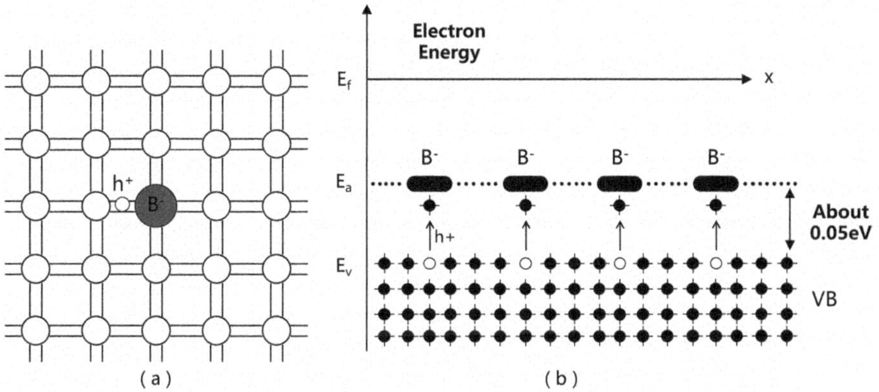


**Fig. 1.4** Structure and band structure of a p-type semiconductor [1]

of holes significantly exceeds the number of intrinsically excited electrons, making holes the majority carriers, while electrons become the minority carriers.

1. **Physical properties of semiconductors**

(1) **Energy-band structure**

A. **Bandgap**

One of the fundamental physical properties of a semiconductor is its energy band structure, which primarily includes the valence band, conduction band, and forbidden band. In this structure, the forbidden band and the conduction band are two crucial concepts. The forbidden band, also known as the energy gap, is the region of energy between the valence band and the conduction band. Electrons cannot exist within this region, which is why it is called the "forbidden band." The width of the forbidden band is an important parameter for characterizing semiconductor materials, as it determines the conductivity of the semiconductor. The larger the energy gap, the lower the conductivity of the material, and vice versa. The conduction band is the energy region where electrons can exist and move freely, allowing electrical conduction. The position and width of the conduction band play a key role in determining the electrical conductivity of the semiconductor. Under certain conditions, when an electron gains sufficient energy, it can transition from the valence band to the conduction band, becoming a free electron and enabling electrical conduction.

B. **Fermi level**

The Fermi level is a crucial concept in the semiconductor band structure, representing the boundary between the electron and hole distributions within the semiconductor. At absolute zero, the electron and hole distributions follow the Fermi–Dirac distribution, with the Fermi level marking the boundary of this distribution. The position of the Fermi level determines the type of semiconductor, such as N-type or P-type.

C. **Electrons and holes**

In semiconductors, electrons and holes are the carriers responsible for conduction. Electrons are fundamental particles with a negative charge. When an electron moves from the valence band to the conduction band, it becomes a free electron and can move freely to conduct electricity. A hole is the absence of an electron in the valence band and can be considered a positively charged particle. When an electron moves from the valence band to the conduction band, a hole is created in the valence band. The concentration of electrons and holes determines the conductivity of the semiconductor.

(2) **Carrier-carrier interactions in semiconductors**

A. **Generation and recombination of electrons and holes**

In semiconductors, charge carriers are particles that can move freely and carry a charge. In pure semiconductors, the charge carriers are mainly composed of electrons and holes.

Electrons are the dominant negative charge carriers in semiconductors, and they can move freely in the band structure of the semiconductor. When the semiconductor is excited by external energy (such as light energy or thermal energy), the electrons in the valence band can obtain enough energy to transition to the conduction band and thus become free electrons. In this process, a vacancy is left in the valence band, namely the hole.

The recombination of electrons and holes means that the free electrons meet and combine with the hole to bring the electron back to the valence band and the holes disappear. During the recombination process, energy is released, a phenomenon called radiation recombination. In addition, there is a non-radiative recombination, in which the energy is released in a non-radiative form (such as phonons).

B. **Mobility and diffusion coefficient**

Mobility is the ability of the carrier to move under an electric field, which is related to the carrier concentration, the temperature, and the intrinsic properties of the material. In semiconductors, the mobility of electrons and holes is affected by various factors, such as impurity concentration, temperature, etc. In general, the higher the mobility, the easier the movement of the charge carrier in the semiconductor.

The diffusion coefficient is the parameter describing the disordered movement of the current carrier in a semiconductor, which is related to the carrier concentration, the temperature, and the diffusion length of the material. The larger the diffusion coefficient, the more intense the disorder movement of the carrier in the semiconductor.

It should be noted that although the mobility and diffusion coefficient are both related to the carrier concentration and temperature of the carriers, their physical significance and influencing factors are different. The mobility mainly reflects the ordered motion of the carrier under the electric field, while the diffusion coefficient reflects the disordered movement of the carrier without the electric field.

### (3) Electrical conductivity of the semiconductor materials

#### A. The relationship between the resistivity and the temperature

The electrical conductivity of a semiconductor is characterized by its resistivity, the extent to which the current flow is hindered inside a material. For semiconductors, their resistivity and temperature are closely related. As the temperature increases, the resistivity of the semiconductor material usually decreases. This is because the temperature rise increases the vibration of the atoms inside the material, hindering the movement of the electrons and thus reducing the resistivity. However, this relationship is not linear, in a certain temperature range, the resistivity change is very sensitive to the temperature, which allows the semiconductor to maintain good conductivity in the environment with large temperature changes.

#### B. Hall effect

Mobility is the ability of a charge carrier to move under the influence of an electric field. It is influenced by factors such as carrier concentration, temperature, and the intrinsic properties of the material. In semiconductors, the mobility of electrons and holes is affected by various factors, including impurity concentration and temperature. Generally, the higher the mobility, the easier it is for the charge carrier to move within the semiconductor [2–4].

The diffusion coefficient is a parameter that describes the random motion of charge carriers in a semiconductor. It is related to carrier concentration, temperature, and the diffusion length of the material. The larger the diffusion coefficient, the more intense the random movement of the carriers in the semiconductor.

It is important to note that while both mobility and the diffusion coefficient are influenced by carrier concentration and temperature, their physical significance and the factors that affect them are different. Mobility primarily reflects the ordered motion of carriers under an electric field, whereas the diffusion coefficient describes the random movement of carriers in the absence of an electric field.

## 2. Semiconductor material classification

### (1) Elemental semiconductor

Elemental semiconductors form the foundation of semiconductor materials, with Si and Ge being the most commonly used. Si has a stable crystal structure, a high melting point, and excellent electrical properties, making it one of the most important materials in the semiconductor industry. Its conductivity lies between that of a conductor and an insulator, allowing it to effectively control the flow of current. Si is widely used in the manufacture of integrated circuits, computer chips, solar cells, and more. Ge, while similar to Si, has higher conductivity and is more stable at higher temperatures. It is often used in high-frequency amplifiers and certain specialized electronic devices. However, Ge is gradually being replaced by Si due to its higher cost.

(2) **Compound semiconductor**

A. **Group III-V compounds**

III-V compound semiconductors are a class of materials with unique electrical properties, such as gallium arsenide (GaAs). These materials exhibit high thermal stability and excellent electrical characteristics, making them widely used in high-frequency and high-speed electronic devices. GaAs has a direct bandgap and higher electron mobility than Si, which makes it ideal for high-speed electronic applications. For example, GaAs is used in the manufacture of microwave communication devices, lasers, photo detectors, and more.

B. **Family compounds (e.g., ZnO)**

The II-VI family of compound semiconductors, such as zinc oxide (ZnO), possesses broadband gap characteristics, making them ideal for optoelectronic and power electronic applications. The bandgap of ZnO is approximately 3.37 eV, which is larger than that of Si, granting it better transparency in the ultraviolet (UV) region. As a result, ZnO is commonly used to manufacture UV detectors and optoelectronic devices. Additionally, ZnO exhibits excellent piezoelectric properties, making it suitable for the production of piezoelectric devices.

C. **Features and applications**

Elemental and compound semiconductors have unique electrical properties, enabling their wide application in electronic and optoelectronic devices. Elemental semiconductors, such as Si and Ge, are commonly used in integrated circuits, computer chips, solar cells, and other fields due to their stable electrical characteristics. Compound semiconductors, such as III-V family compounds, are crucial in high-frequency, high-speed electronic devices, optoelectronic devices, and power electronics. Their special electrical properties and broadband bandgaps make them ideal for these applications.

(3) **Organic semiconductors**

A. **Structure and conduction mechanism**

Organic semiconductors are primarily organic compounds composed of carbon, hydrogen, nitrogen, sulfur, and other elements. Their structures are highly diverse, and different properties can be achieved through molecular design. The conduction mechanism of organic semiconductor materials mainly involves the migration of $\pi$ electrons within the organic conjugate system under an external electric field. When subjected to an electric field, the $\pi$ electrons in the conjugated system can migrate between molecules, generating an electric current. This mechanism allows organic semiconductors to exhibit semiconductor properties under certain conditions.

The electrical conductivity of organic semiconductors is closely related to their structure. Generally, the higher the degree of conjugation and the longer the molecular chain, the better the electrical conductivity. Additionally, factors such as substituent groups in the molecular structure and intermolecular interactions also influence the electrical conductivity of organic semiconductors.

B. **Application in flexible electronics and wearable devices**

Due to the unique properties of organic semiconductors, such as good flexibility, solution processability, and low cost, they have broad applications in flexible electronics and wearable devices. In flexible electronics, organic semiconductors can be used to manufacture flexible circuits, flexible displays, flexible sensors, and so on. For example, flexible circuits can be applied to mobile phones, flexible displays can be applied to foldable mobile phones, flexible sensors can be applied to health monitoring devices, etc. For wearable devices, organic semiconductors can be used to make flexible solar cells, flexible energy storage devices, biosensors, etc. For example, flexible solar cells can be applied to wearable power supply, flexible energy storage devices can be applied to data storage in wearable electronic products, and biosensors can be applied to monitor the physiological parameters of the human body in real-time.

3. **Semiconductor doping and processes**

(1) **Doping technology**

A. **N-type doping and P-type doping**

N-type and P-type doping are two basic doping techniques in semiconductor fabrication. The purpose of doping is to change the conductance properties of semiconducting materials by introducing external atoms.

(a) **N-type doping**: In N-type semiconductors, pentavalent elements such as phosphorus (P), arsenic (As), or antimony (Sb) replace some of the Si atoms in the lattice. These elements have a larger atomic radius than Si, which alters the energy band structure of the Si lattice. The introduction of trivalent impurity atoms creates free electrons in the semiconductor, which act as charge carriers. As a result, N-type semiconductors have an excess of free electrons, giving them a negative overall electrical charge.
(b) **P-type doping**: P-type semiconductors are created by incorporating trivalent elements such as boron (B), aluminum (Al), or gallium (Ga). These elements have smaller atomic radii than Si, causing vacancies to form in the Si lattice, which can capture free electrons. Since trivalent impurity atoms have only three valence electrons, they do not fully bond with the four valence electrons of Si, leaving behind "holes." These holes can act as charge carriers, which gives P-type semiconductors their positive electrical activity.

E. **Effect of doping on the semiconductor electrical conductivity**

Doping has a significant influence on the electrical conductivity of semiconductors, which is mainly reflected in the following aspects:

(a) **The increase of charge carriers**: Doping will introduce additional charge carriers (free electrons or holes), which can flow under the action of the electric field, thus increasing the conductivity of the material.

(b) **Adjustment of the energy band structure**: Doping will change the energy band structure of the semiconductor. For example, N-type doping will form an additional conduction band above the valence band, while P-type doping will form an additional valence band below the conduction band. This band adjustment directly affects the electron transition and the charge movement.

(c) **Type conversion**: Through doping, the original N-type or P-type semiconductor can be transformed into another type or a uniform N-type or P-type semiconductor can be prepared, which is crucial for the construction of various electronic devices.

Doping is one of the key technologies in the semiconductor process. By accurately controlling the type and concentration of doping, semiconductor devices can be designed to meet specific requirements, which can be applied in many fields such as microelectronics, optoelectronics, and power electronics.

(4) **Semiconductor manufacturing processes**

A. **Crystal growth and cutting**

Crystal growth is the first step in semiconductor manufacturing, and the most common method used is the Czochralski process (CZ growth). In this method, molten semiconductor material, such as Si, is poured into a container called a crucible. The bottom of the crucible is then gradually lifted while being rotated at high speed, causing the molten material to move upward along the rotating axis. As the molten material cools, it solidifies on the top of the crucible, forming a cylindrical crystal. The growth and rotation rates of the crystal can be controlled to produce crystals of varying diameters.

Once the crystal is grown, it must be sliced into thin sheets, known as Si wafers. This process is typically performed using a multi-wire cutter, which employs filaments (such as diamond wires) that apply force while rotating at high speeds to cut the crystal. The resulting Si wafers are extremely thin, usually only a few microns thick.

B. **Preparation and cleaning of Si wafer**

After the Si wafer is cut, it needs to be prepared and cleaned to remove the damage and contaminants generated during the cutting process. First, the Si wafer will be polished to remove scratches and irregularities from the surface. Polishing is usually achieved using chemical corrosion and mechanical grinding. The Si wafer then goes through a series of cleaning steps, including the use of acid and alkali solutions to remove organic and inorganic surfaces, and washing with deionized water. These cleaning steps ensure that the surface of the Si wafer is smooth and pollution-free, and they are ready for subsequent process steps.

C. **Oxidation, lithography, etching**

After the preparation and cleaning of the Si wafer, the oxidation, lithography, etching, and other process steps can be carried out.

Oxidation is a key step in semiconductor manufacturing, used to form an insulating layer of silica on the surface of a Si wafer ($SiO_2$) layer. This process is usually carried out at high temperatures, using oxygen or ozone as an oxidant. $SiO_2$ layers can be used as insulating layers for use to isolate and protect different parts of the circuit.

Photography is a technique used to create patterns on a Si wafer. It involves using photosensitive chemicals (photoresists) and light sources to transfer circuit patterns onto the wafer. First, the surface of the Si wafer is coated with a layer of photoresist. Then, it is exposed to a light source (such as ultraviolet light), causing chemical changes in the exposed areas of the photoresist. Finally, the circuit pattern is transferred to the Si wafer by removing the unexposed photoresist.

Etching is a technique used to remove material from specific areas on a Si wafer. It employs chemical solutions or plasma, depending on the patterns created during the lithography process, to etch away the unprotected areas. Etching allows for precise circuit processing to form the desired fine structures.

The detailed operation and control of these process steps are crucial, as they directly affect the performance and reliability of semiconductor devices. The precision of factors such as temperature, pressure, and time must be carefully controlled throughout the manufacturing process to ensure the quality and performance of the final semiconductor devices.

### (5) **Preparation of semiconductor devices**

#### A. **Fabrication and characteristics of PN junction**

The PN junction is the foundation of semiconductor devices, and its fabrication is primarily achieved through doping. Doping is categorized into n-type and p-type, where n-type semiconductors are doped with penta-valent elements, such as phosphorus (P), and p-type semiconductors are doped with trivalent elements, such as boron (B). In semiconductor materials like Si or Ge, PN junctions are formed by carefully controlling the doping concentration and the doping regions.

When the n-type and p-type semiconductors come into contact, a depletion zone is formed at the junction, which contains almost no free carriers, and is referred to as the space charge zone. The PN junction exhibits a unidirectional conductivity characteristic: under forward bias, electrons and holes easily cross the junction, allowing current to flow, while under reverse bias, almost no current flows.

#### B. **Manufacturing of diodes, transistors, and other devices**

The manufacture of diodes is based on the characteristics of PN junctions, featuring a main structure composed of alternating layers of P-type and N-type semiconductors. During the manufacturing process, the PN junction is initially formed through doping. Subsequently, technologies such as epitaxial growth, ion implantation, or molecular beam epitaxy are employed to grow one or more layers of semiconductor material, completing the diode's structure. Finally, metallization is performed to connect electrodes to both ends of the PN junction, enabling the diode's conductive function.

The manufacturing process of transistors is more complex and involves three main regions: the emitter, base, and collector. First, the PN junction is formed through doping. Then, ion implantation or molecular beam epitaxy is used to grow a layer of semiconductor material, establishing the transistor's basic structure. The emitter and collector regions are subsequently created on either side of the base region, with the emitter region doped with a high concentration of impurities and the collector region doped with a lower concentration. Finally, metallization is carried out to connect electrodes to the emitter, base, and collector regions, enabling the transistor's control and amplification functions.

Doping and process control are critical factors in semiconductor device fabrication, as they determine the performance and reliability of the devices. By continuously optimizing doping techniques and manufacturing processes, the performance of semiconductor devices can be enhanced, fostering the advancement of the electronics industry.

4. **Semiconductor devices and their applications**

(1) **Diode**

A. **Working mechanisms and characteristics**

A diode is a semiconductor device whose core component is a PN junction, formed by P-type and N-type semiconductors. When the P terminal is at a positive voltage and the N terminal is at a negative voltage, the diode operates in a forward-biased state. Conversely, when the P terminal is at a negative voltage and the N terminal is at a positive voltage, the diode is in a reverse-biased state. In the forward-biased state, the diode's forward voltage exceeds its conduction voltage (approximately 0.7 V for a Si diode and around 0.3 V for a Ge diode), allowing current to flow through the diode. In the reverse-biased state, the reverse voltage is below the cutoff voltage, preventing current from flowing through the diode. The primary characteristics of a diode include its forward and reverse characteristics, temperature characteristics, and dynamic response characteristics, among others.

B. **Application of rectification and stabilization**

Diodes are primarily used in circuits for applications such as rectification, voltage stabilization, modulation, and protection. In rectification applications, diodes convert AC electrical signals into pulsating DC signals, including full-wave and half-wave rectification. In voltage stabilization applications, diodes serve as voltage regulation elements, such as Zener diodes and voltage regulator diodes. The working principle is that when the voltage exceeds its rated value, the diode conducts quickly, diverting the excess voltage and maintaining stability in the circuit. Diodes are also employed in modulation circuits, such as amplitude modulation and frequency modulation, and in protection circuits, including overvoltage protection and reverse polarity protection.

(2) **The transistors**

A. **Field-effect transistors (FET) and bipolar junction transistors (BJT)**

Field-effect transistors (FETs) and bipolar junction transistors (BJTs) [5–12] are two common types of transistors, differing primarily in their input characteristics and working principles.

The input characteristic of an FET is voltage-controlled, meaning the input signal controls the drain current by altering the gate voltage. In contrast, the input characteristic of a BJT is current-controlled, where the input signal regulates the collector current by changing the base current. An FET conducts when the gate voltage exceeds the threshold voltage and enters cutoff when the gate voltage is below the threshold voltage. A BJT operates when the base current exceeds the cutoff current, and it turns off (cuts) when the base current is lower than the cutoff current.

B. **Amplification, and switching**

Transistors are widely used in circuits for applications such as amplification, switching, voltage stabilization, and oscillation. In amplification applications, transistors can amplify the input signal by a certain factor, such as in amplifier cascades or power amplification. In switching applications, transistors enable high-speed switching operations, such as logic gates in digital circuits and switches in power management circuits. Transistors are also used in voltage regulator circuits, including voltage regulator tubes and transistor-based regulators. The working principle involves controlling the conduction and cutoff of transistors to maintain a stable voltage within the circuit. Additionally, transistors are employed in oscillating circuits, such as LC oscillators and RC oscillators, as well as in analog signal processing circuits, including filters and amplifiers.

(3) **Integrated circuits**

An integrated circuit (IC) is a compact assembly of numerous small electronic components such as transistors, resistors, capacitors, and other elements, manufactured using semiconductor technology. These components are integrated into a small Si wafer to perform specific electronic functions. Integrated circuits are the cornerstone of modern electronic technology, significantly advancing miniaturization, performance, and cost reduction.

Based on different manufacturing processes, integrated circuits can be categorized into small-scale integrated circuits (SSI), medium-scale integrated circuits (MSI), large-scale integrated circuits (LSI), and very-large-scale integrated circuits (VLSI). Integrated circuits can also be classified according to their functions into analog integrated circuits, digital integrated circuits, and hybrid integrated circuits. Analog ICs are used to process continuous signals, such as sound and light intensity, while digital ICs handle signals expressed in digital form, such as data processing in computers. Hybrid integrated circuits contain both analog and digital components.

The manufacturing process of integrated circuits includes the following stages: design, mask fabrication, wafer fabrication, packaging, and testing. During the design stage, electronic design automation (EDA) tools are used for circuit design and simulation. Mask fabrication involves transferring the designed circuit pattern onto the Si wafer. Wafer manufacturing involves growing pure Si single crystals using the Czochralski method or other techniques. Wafer processing consists of steps like doping, lithography, etching, deposition, and ion implantation to form the circuit. Packaging is the process of encasing the processed wafer into usable electronic components. Finally, testing ensures that the integrated circuit functions meet the design specifications.

(4) **Applications of semiconductors**

A. **Solar energy cells**

Solar cells [13–15] are devices that utilize semiconductor materials, such as Si and Ge, to directly convert sunlight into electricity. Si solar cells are the most widely used type. When sunlight strikes a solar cell, the semiconductor material absorbs the light energy, generating electron–hole pairs. Through an external circuit, the electrons are transferred, producing an electric current. The efficiency of solar cells is influenced by factors such as the properties of the materials used, the cell's structure, and the light conditions. Currently, high-efficiency solar cells can achieve efficiencies exceeding 20%, while multi-junction solar cells can reach efficiencies of over 40%. Solar cells have a wide range of applications, including powering space satellites, telecommunications systems, rural electrification projects, and solar chargers.

B. **LED**

LED is a semiconductor device that can directly convert electrical energy into light energy. It consists of n-type and p-type semiconductors, and the PN junction formed when the two semiconductors are in contact will emit light after energizing. LED has the advantages of high efficiency, low consumption, long life, rich color, and so on, and is a new generation of lighting light source. LED lighting has been very widely used, from simple indicators, and display screens to indoor and outdoor lighting, car taillights, medical equipment, and so on have LED figure. LED lighting not only saves energy and reduces carbon emissions, but also provides better lighting effects and more flexible design possibilities.

C. **Sensor technology**

A sensor is a device that can feel non-electrical physical quantities (such as temperature, pressure, humidity, light intensity, etc.) and convert them into electrical signals. Semiconductor sensors are one of them that use the specific properties of the semiconductor material to detect and convert signals. For example, the temperature sensor is fabricated by using the characteristics of the resistance of the semiconductor material. Sensors are widely used in industrial automation, medical equipment, automotive electronics, environmental monitoring, and other fields. They can be used to monitor the running status of the equipment, realize intelligent control, improve measurement accuracy, and so on.

D. **Semiconductor laser**

A semiconductor laser is a semiconductor device capable of generating a laser, which is an indispensable technology in modern communication, data storage, laser printing, medical surgery, and other fields. The semiconductor laser can produce a laser in the semiconductor active region by injecting the current, which has the advantages of small size, lightweight, high electro-optical conversion efficiency, and fast working speed. Semiconductor lasers can be divided into infrared lasers, visible lasers, ultraviolet lasers, and so on. They are widely used in optical communication, optical storage, laser printing, lidar, and other fields.

6. **Development bottleneck and countermeasures of China's third-generation semiconductor technology**

(1) **Analysis of the development status and trend of the third-generation semiconductor technology in China**

A. **The third-generation semiconductor market is in a rapid growth stage**

As a strategic emerging industry and an important starting point to achieve energy conservation and emission reduction, the third generation of semiconductor technology and industry has been highly valued by the national and local governments, as well as enterprises. At present, China has formed a relatively complete industrial system and strong technology research and development ability in this field and has gradually given birth to some innovation teams with industrial core competitiveness. Compared with the first and second-generation semiconductors, China the United States, and other countries are in the early stage of the outbreak of the third-generation semiconductor technology industry, and are facing a good opportunity to turn over against the trend. In the next 10 to 15 years, China will have a great opportunity to develop from a third-generation semiconductor industry power to an industrial power. According to CASA data, the market size of China's third-generation semiconductor reached 11.179 billion yuan in 2022, with a year-on-year growth of 39.2%. The compound growth rate from 2018 to 2022 was 43%, with a fast growth rate. In 2022, the market size of gallium nitride (GaN) semiconductors reached 6.258 billion yuan, the market size of Si carbide (SiC) semiconductors reached 4.345 billion yuan, and other compound semiconductors reached 576 million yuan.

(2) **Enterprises in the main links of the industrial chain have emerged one after another**

In recent years, China's third generation of semiconductors has initially formed five key development regions, namely, Beijing-Tianjin-Hebei Shandong, Yangtze River Delta, Pearl River Delta, Fujian Delta, and central and western regions. Beijing and Guangdong have successively established the Beijing third-generation semiconductor Materials and Application Joint Innovation base and the third-generation semiconductor South base. Jiangsu, Zhejiang, Xiamen, Jiangxi, and other places have gradually become the third-generation semiconductor industry characteristic agglomeration area with a relatively complete industrial chain. Among them, the

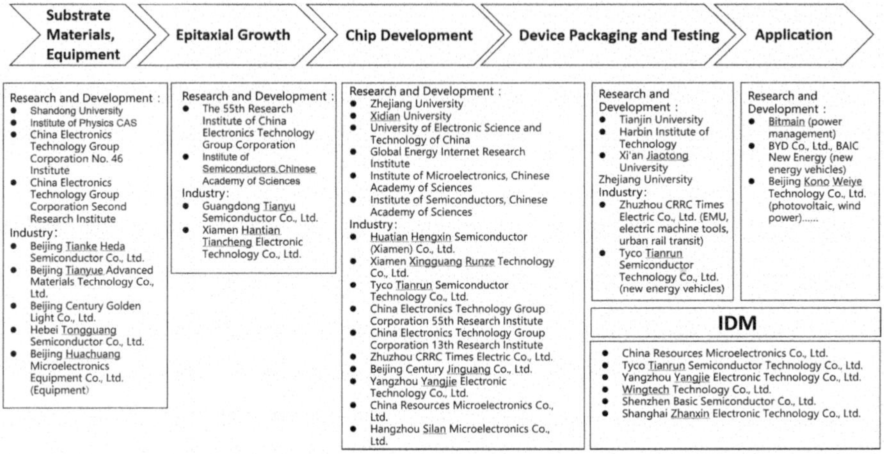


**Fig. 1.5** Major enterprises in all links of the SiC power electronic device industry chain in China

representative enterprises of the third generation semiconductor in Jiangsu province are the most distributed, such as Suzhou Ni Technology Co., LTD. (substrate), Innosecco Technology Co., LTD. (extension), Suzhou Nengxun High Energy Semiconductor Co., LTD. (design), Yangzhou Yangjie Electronic Technology Co., LTD. (manufacturing, sealing, and testing). From the perspective of the industrial chain, each link of SiC power electronic devices is SiC single crystal substrate-SiC material epitaxy-device design-device manufacturing-packaging and testing-module. At present, China has a complete industry chain of SiC power electronic devices, as shown in Fig. 1.5.

In the field of Si carbide wafers, although the market share of Si carbide wafers in the United States is more than 60%, the domestic Shandong Tianyue Advanced Material Technology Co., Ltd. and Beijing Tiankeheda Semiconductor Co., Ltd. have emerged. Semi-insulating Si carbide substrate, the global market concentration is high. Public data show that in 2020, Wolf speed, II-IV, and Shandong Tianyue in the United States are the three dominant, accounting for 98%, among which Shandong Tianyue occupies 30% of the market share. In terms of conductive Si carbide substrate, Wolfspeed in the United States has an absolute monopoly position, with a market share of as high as 62% in 2020. II-VI, SiCrystal, SKSiltron, Tiankeheda, and other enterprises occupy the remaining market, among which a market share of 4%.

In terms of extension, the gap between China and the international leading level is narrowing. Domestic enterprises already supply 6 ~ 8 inches of extension, but the production quality needs to be further improved. Europe has a big advantage in SiC epitaxy tablets, with famous companies including Norstel in Sweden, Infineon in Germany, and STMicroelectronics in Switzerland. According to public data, China's Dongguan Tianyu and Hantiantian can supply 3 inches to 6 inches of SiC extension, and domestic enterprises have been able to mass-produce 8-inch Si-based GaN epitaxial wafers.

In terms of device manufacturing, the gap between China's patent reserve and the international leading level is relatively small. According to the data of the Orbit patent database, as of March 2021, there are more than 20,000 patents related to GaN-based FET devices worldwide, of which the United States accounts for 31.1%, Japan for 22.2% and China for 21.0%. Wolf-speed, TSMC, and Xidian University have certain advantages in this field. Micro LED There are more than 2500 related patents worldwide, of which China accounts for 40.8%, the United States for 25.7% and Japan for 6.1%.

Typical domestic enterprises include SAN'an Optoelectronics, Wentai Technology and so on. In June 2020, Wentai Technology completed the acquisition of Anshi Semiconductor of the Netherlands. In 2018, the total production volume of Anshi Semiconductor exceeded 100 billion units, ranking first in the world. In 2019, Anshi Semiconductor was the first in the world to deliver GaN gallium nitride semiconductor GaN-FET, becoming the only compound power semiconductor company in the industry to mass produce and deliver to customers. SAN'an Optoelectronics has made great progress in Si carbide products, realizing the quasi-mass production of 8-inch substrates, and some products have entered the supply chain of mainstream new energy vehicle enterprises. In the first half of 2023, the SiC market demand was strong, and the overall performance of Hunan SAN'an maintained a growth trend, achieving a revenue of 582 million yuan, with a year-on-year growth of 178.86%. Its strategic cooperation with ST STMicroelectronics will have a positive impact on the SiC product penetration rate, revenue scale, and profitability of SAN'an Optoelectronics. In addition, the 55th Research Institute of China Electronics Technology Group Corporation and the 13th Research Institute of China Electronics Technology Group Corporation have both achieved outstanding performance of power devices.

(3) **The third-generation semiconductor industry will continue to maintain rapid growth**

As an important part of the National Science and Technology Innovation 2030 major project "RESEARCH and application of key new materials", the development of third-generation semiconductor technology has been widely valued [16, 17]. The third-generation semiconductor is an important development direction to help society save energy conservation and emission reduction and achieve the goal of "carbon neutrality". Shunyi District of Beijing, Jinan City of Shandong Province, and Pingshan District of Shenzhen City are divided into the key areas of the third-generation semiconductor industry, which are the three major agglomeration areas with the significant development of the third-generation semiconductor industry in China.

China has a good foundation for third-generation semiconductor technology research and development and has a large number of intellectual property rights. Tsinghua University, Peking University, Chinese Academy of Sciences, Nanjing University, Shandong University, Xi'an University of Electronic Science and Technology, Sun Yat-Sen University and other university research institutions and China Electronics technology group research in more than 20 years, and through technology transformation and cooperation established the gallium semiconductor company, Suzhou Technology Co. LTD., Shandong Day Yue advanced Technology Co., LTD., Beijing Day or Da Semiconductor Co., LTD., Hebei with light Semiconductor Co., LTD., Dongguan Day Domain Semiconductor Technology co., LTD., and other enterprises. With the development prospects of the third-generation semiconductor in 5G, Internet of Things, energy, and other fields becoming increasingly clear, industrial development will continue to accelerate, the demand for talent and technology is increasing, and the advantages of China's scientific research institutions in talent and technology reserve will become more and more significant.

China is the main sales market of the world's third-generation semiconductor leading enterprises. It is expected that in terms of the future market size, China's third-generation semiconductor industry will continue to maintain rapid growth. In terms of segmentation product development trend, SiC demand will increase, and GaN [18] application scenarios will be further expanded (Table 1.1).

(4) **The bottleneck facing the development of China's third-generation semiconductor technology**

A. **Lack of top-level design, and the lack of original innovation capability and talent pool**

China has been developing third-generation semiconductor research and development since the early twenty-first century. The National Natural Science Foundation of China and the National Key Research and Development Program have launched

**Table 1.1** Comparison of the mastery degree of the third-generation semiconductor core technologies

| Carborundum | China | External |
|---|---|---|
| Underlayment | Mainly in 4 inches, gradually transition to 6 inches | Mainly at 6 inches, and is transitioning to 8 inches |
| Extension | Mainly in 4 inches, with a small number of 6 inches epitaxial tablets. But the cost is high and the quality is unstable | Mainly in 4 to 6 inches. High yield and good quality control |
| Device | Mass production has been achieved | Mass production has been achieved |
| Gallium nitride | China | External |
| Underlayment | It has achieved small batch production of 2 inches and 4-inch substrate | Has achieved 2–4 inches of mass production, 6-inch products have realized commercial use |
| Extension | Basically, in line with the international advanced level | |
| Device | Microwave power devices and MMIC products, GaN RF devices have been mass produced; photoelectric devices LED efficiency above 180 lm/W | The highest voltage of Si GaNHEMT of electronic devices is 650 V; the maximum power of RF device is 1800 W |

related research projects, resulting in a series of achievements. However, despite the widespread application of third-generation semiconductor technology, China's innovation capacity in this field remains insufficient. Several significant scientific and technological challenges remain, including issues such as P-type doping in third-generation semiconductors for deep ultraviolet light and laser applications, as well as point defects in these materials (Zhang Liangliang, Lee, 2021).

From a systemic and structural perspective, China's support for third-generation semiconductor technology has largely been focused at the national strategic level. In contrast, the U.S. CHIPS and Science Act of 2022 outlines detailed R&D investment plans and talent development programs, reflecting a more systematic and strategic approach from the government. In terms of R&D investment, China's support for third-generation semiconductor research is lower compared to developed countries that have long-term, stable investments in the sector. China's efforts remain fragmented, with limited and inconsistent funding, making it difficult to address foundational research and key technological challenges. Moreover, from a talent perspective, the lack of professionals specialized in third-generation semiconductors is a core factor limiting China's innovation capabilities. The country has yet to establish a systematic training system for these specialists, and recruitment is further constrained by the international situation.

B. **Key production equipment is in short supply, and the manufacturing technology is not yet fully mature**

The lack of key production equipment in China's semiconductor sector has been a long-standing issue. In chip production, the lithography machine is crucial to the manufacturing process. In recent years, countries like the United States and Japan have imposed export restrictions on advanced chip manufacturing equipment to China, including a blockade on lithography machines. At present, China's domestically developed technology and equipment cannot fully meet the needs of industrial development, which further hampers the advancement of high-end chips in the country.

China's core manufacturing technology for third-generation semiconductors is still not fully mature. Issues such as stability and reliability remain significant weaknesses, posing large production risks for enterprises. For example, the production of large-size wafers requires advanced material technology and production techniques. However, China's domestic large-size wafer manufacturing technology is still in its early stages, characterized by high costs and low yields. Despite this, domestic companies are aggressively deploying large-size wafer production lines, which increases production risks and leads to issues with redundant construction. According to statistics, by the end of 2023, China will have at least 86-inch SiC wafer production lines, with approximately 10 SiC production lines under construction. Additionally, 54-inch GaN-on-SiC production lines and five GaN RF production lines are currently being built.

C. **High cost of the pilot platform and low level of engineering technology**

The research, development, and industrialization of third-generation semiconductors require expensive production and process equipment, high-grade clean environments, and advanced testing and analysis platforms, all of which involve substantial capital investment and high operational costs. Currently, most domestic research institutions and enterprises involved in third-generation semiconductor development operate individually. There are four 6-inch SiC device pilot lines, including those at CRRC Times, Century Golden Light, the Global Energy Internet Research Institute, and the 55th Research Institute of China Electronics Technology Group Corporation. However, due to the lack of open public research and development, service, and industrial pilot platforms, many enterprises struggle to conduct engineering research, development, and testing. As a result, engineering technology has become a significant bottleneck, slowing the pace of research and development innovation in the third-generation semiconductor sector.

D. **Domestic materials and devices are difficult to enter the application supply chain, and the technology industrialization capacity is weak**

Downstream applications of third-generation semiconductors span nearly all areas of electricity consumption, including optoelectronics, power electronics, and microwave radio frequency technologies. Foreign third-generation semiconductor companies are forming strong partnerships with terminal application enterprises to establish an industrial presence. For example, in 2019, STMicroelectronics was selected by the Renault-Nissan-Mitsubishi Alliance to supply power electronics for the advanced onboard charger (OBC) of their next-generation electric vehicles.

Currently, China's third-generation semiconductor materials and devices (such as MicroLED) are making continuous breakthroughs in key technologies. However, the industrialization of these technologies and their market application have not developed in tandem. Issues such as the lack of coordination between standards, testing, certification, and industry practices, as well as challenges in integrating domestic devices into the application supply chain, contribute to slow development. The lengthy cycles and difficulties in application validation and iterative research and development have resulted in slow progress in industrialization.

(5) **Suggestions on the development of the third-generation semiconductor technology in China**

A. **Increase R&D support and strengthen the innovation capacity of core technologies**

The 14th Five-Year National Key Research and Development Plan has identified third-generation semiconductors as a key development focus. It aims to address product availability, related equipment, and continuous innovation capabilities. It is recommended that the government prioritize key areas when making plans, ensuring overall coordination and long-term strategies. First, the implementation plan for key research and development initiatives should be further refined, with a focus on fully deploying core technologies for third-generation semiconductor materials.

Second, investment and financing institutions, along with enterprises, should be guided to establish dedicated funds for third-generation semiconductors to support fundamental research and breakthroughs in core technologies. This support should particularly target competitive teams and platforms to ensure stability and growth.

B. **Accelerate the deployment of the whole industrial chain and improve the ecology of the industrial system**

The development of the third-generation semiconductor industry is a large-scale, systematic project. Given the current stage of technological development and industrial maturity in China, we will accelerate the deployment of the entire industrial chain. Through government involvement, we will strengthen upstream and downstream cooperation within the industry, fostering a healthy industrial ecosystem. Leveraging relevant national platforms, we will integrate the research and development capabilities of enterprises, universities, and research institutes to enhance manufacturing technology and address challenges in the generic technologies needed for industrialization. We will establish a collaborative community that spans the entire chain—research and development, manufacturing, industrialization, and application—and promote the sustainable ecological development of the industrial system.

C. **Promote development through application and accelerate the transformation of technological achievements**

China's new energy industry has gradually become the global "leader", promote the acceleration of industrial chain upgrade, product rapid iteration and technology innovation, the domestic market demand for the third generation of semiconductor materials and devices is rapidly increasing, terminal application enterprises constantly adjust the supply chain, to local semiconductor materials enterprise provides a rare opportunity for development. With the help of the capital entry vigorously advocated by the independence of the semiconductor industry, it is suggested that the local government can establish the pilot demonstration organization and implementation mechanism of "the integration of technology supply and market drive", accelerate the transformation and industrialization of technological achievements, and promote the application of localization.

D. **Strengthen international cooperation in science and technology and enhance international competitiveness**

We will establish precise, in-depth, and various forms of scientific and technological cooperation. Strengthen in-depth cooperation between national institutions, enterprises, and R&D teams. Actively participate in the formulation of international standards, and gradually enhance the right to speak. For countries with relatively mature industrial chains and advantages, we should focus on cooperating with enterprises and organizations with mature technologies to attract excellent talents and build a talent system of key talent teams.

In recent years, with the dual support of policy and capital, the scale and technology level of China's semiconductor industry has been significantly improved, and the industrial structure has been further optimized. Semiconductor enterprises are ushering in a valuable time window opportunity, and industry leaders have launched large-scale investments to win the development opportunity. At present, nearly 100 units in China have carried out research on the third generation of semiconductor materials and applications, made breakthrough progress in basic science issues, accumulated certain industrialization experience, and trained a large number of scientific research talents and innovation teams. At present, the local industrial chain layout of China's semiconductors is accelerating, and the third-generation semiconductor industry will usher in new development opportunities.

## 1.3 Basic Physical Mechanisms of Photoelectric Detection Devices

1. **Basic principle of photoelectric effect**

(1) **Overview of the photoelectric effect**

The photoelectric effect occurs when light strikes a metal surface. If the frequency of the light is greater than the threshold frequency of the metal, electrons are emitted from the surface. These emitted electrons are called photoelectrons. The photoelectric effect is one of the fundamental experimental phenomena in quantum physics, demonstrating the particle-like nature of light. According to Einstein's quantum theory of light, light can be described as a stream of particles called photons. When a photon interacts with an electron on the metal surface, and the photon's energy is greater than or equal to the minimum energy required for the electron to escape from the metal surface (known as the work function), the electron is excited and emitted from the surface.

(2) **External photoelectric effect**

The external photoelectric effect refers to the movement of escaping photoelectrons in the space outside the metal surface during the photoelectric effect. When photoelectrons are emitted from the metal surface, they are accelerated by the electromagnetic field, resulting in a charge distribution outside the metal surface. This charge distribution generates an electric field known as the optoelectronic electric field. Experimental observations of the external photoelectric effect reveal that the kinetic energy of photoelectrons is independent of the intensity of light and depends solely on the frequency of the light.

(3) **Internal photoelectric effect**

The internal photoelectric effect refers to the movement of photoelectrons within the metal during the photoelectric effect. When a photon is absorbed, it generates an electron–hole pair inside the metal. These electron–hole pairs move within the metal and gradually migrate toward the metal surface. During this process, some electrons are trapped by impurity atoms or defects, leading to a charge distribution inside the metal. This charge distribution generates an electric field, known as the internal photoelectric effect electric field. Experimental results of the internal photoelectric effect indicate that the kinetic energy of photoelectrons depends on both the intensity and the frequency of the light.

A. **Photoconductivity effect**

The photoconductivity effect refers to the increase in electrical conductivity of a material when light strikes the surface of a semiconductor or an insulator. This occurs because light generates electron–hole pairs within the material, which increases the concentration of charge carriers and enhances electrical conductivity. The photoconductivity effect forms the basis of modern photo detectors.

B. **Photovoltaic effect**

The photovoltaic effect refers to the generation of an electromotive force (EMF) on the surface of a semiconductor when light strikes it. This occurs because, under light irradiation, an electron–hole pair is created within the semiconductor, forming a charge distribution on the surface. This charge distribution generates an electric field, which in turn creates an electromotive force on the semiconductor surface. The photovoltaic effect forms the basis of solar cells and other photovoltaic devices.

(4) **Photothermoelectric effect**

The photothermoelectric effect refers to the generation of heat within a material when light strikes its surface, leading to changes in the material's electrical properties. This effect often manifests as a change in the material's resistance, known as the photothermal resistance effect. The photothermoelectric effect is widely used in the fields of thermoelectric devices and photothermal conversion.

2. **Physical mechanisms for photoelectric detection devices**

(1) **Photoemission effect**

The photoemission effect [19] is an important physical phenomenon in photoelectric detection devices, describing the escape of electrons from the surface of a material when irradiated by photons. The following is a detailed analysis of the conditions, processes, quantum theoretical explanations, and applications involved.

A. **Conditions of photoemission**

The occurrence of photoemission effect needs to meet two basic conditions: first, the photon energy must be greater than the escape work (or work function), that is, the energy carried by the photon must be large enough to overcome the surface potential barrier of the material and make the electron escape; second, the frequency or wavelength of the photon must meet specific conditions, because the absorption and response frequency of light of different materials are different.

The process of photoemission can be roughly divided into three steps: first, the photon interacts with the electrons in the material, and the photon energy is absorbed by the electrons; second, the electrons that absorb the photon energy escape from atoms or molecules to form free electrons; finally, these free electrons are collected or detected under the action of the electric field.

B. **Quantum theoretical explanation of photoemission**

The photoemission effect can be explained by quantum theory. According to quantum theory, light is composed of particles called photons, each carrying a specific amount of energy. When a photon interacts with the electrons in a material, the photon energy is absorbed by the electrons, causing them to transition from a lower to a higher energy level. If the photon energy is sufficient to exceed the work function of the material, the electron can escape from the surface of the material. This is the quantum theoretical explanation of photoemission.

C. **Application of photoemission devices**

The photoemission effect is widely used in optoelectronic detection devices. Among them, the most typical phototransmitter devices are phototube and photo multiplier tubes (PMTs). The phototube uses the photoelectric emission effect to convert the optical signal into the electrical signal and realize the photoelectric conversion. The photo-multiplier multiplier amplifies the weak optical signal into a strong electrical signal through the multistage multiplication process, which improves the sensitivity and accuracy of photoelectric detection.

In addition to photoelectric tubes and PMTs, the photoelectric emission effect is also widely used in solar cells, photo detectors, photosensors, and other fields. For example, in a solar cell, the photoemission effect is a key step in realizing the conversion of light energy into electric energy; in a photo detector, the photoemission effect is used to detect information such as energy and frequency of photons; and in a photosensor, to realize applications such as non-contact measurement and detection.

The photoelectric emission effect is an important physical phenomenon in photoelectric detection devices, which not only reveals the basic law of the interaction between light and matter but also provides an important theoretical basis and technical support for the development and application of photoelectric detection technology.

(2) **Photoconductivity effect**

A. **Mechanism of the photoconductivity effect**

The photoconductivity effect [20] is the phenomenon that the electrical conductivity (or resistivity) of certain substances changes when the light shines on the surface. The mechanism of this phenomenon mainly involves the interaction of photons and electrons in matter.

Specifically, when a photon (a particle of light) incident on the surface of matter, if the energy of the photon is large enough (greater than or equal to the binding energy of electrons in the matter), it can be absorbed by the electrons in the matter. Electrons that absorb the photon energy will obtain enough kinetic energy to become free from the nucleus and become free electrons. These free electrons move inside the matter, increasing the concentration of the carriers inside the matter, thus changing the conductivity of the matter.

The photoconductivity effect can be divided into the internal photoconductivity effect and the external photoconductivity effect. The inner photoconductivity effect refers to the change of conductivity when light shines into the substance, while the external photoelectric conductivity effect refers to the phenomenon that the electrons near the surface escape into vacuum when light reaches the surface of the substance, forming photoelectron emission, and then producing the change of conductivity.

B. **Characteristics and applications of photoconductive devices**

A photoconductive device is an electronic device made of photoconductive effect, which has the following characteristics: (1) high sensitivity: because the photoconductive effect is the result of the direct interaction between photons and electrons, the photoconductive device is very sensitive to light and can detect weak light signal. (2) Fast response speed: the response time of photoelectric conductive devices is usually very short and can reach nanoseconds or even microsecond levels, so it is suitable for high-speed photoelectric detection and signal processing. (3) Wide spectral response range: different photoconductive materials have different response wavelength ranges to light, so appropriate materials can be selected to realize the detection of light signals at different wavelengths. (4) Can be made into a variety of shapes and sizes: photoconductive devices can be made into a variety of shapes and sizes according to the needs, and convenient integration into various systems.

Photoconductive devices have a wide range of applications in various fields, such as (1) optical communication: Photoconductive devices can be used for optical reception and optical amplification in the optical communication system, to improve the performance and reliability of the optical communication system. (2) Photoelectric detection: Photoelectric conductivity devices can be used in various photoelectric detection equipment, such as infrared detectors, ultraviolet detectors, etc., to realize the detection and identification of different wavelengths of optical signals. (3) Imaging technology: Photoconductive devices can be used in imaging devices, such as charge-coupled devices (CCD) [21–37] and complementary metal oxide semiconductor (CMOS) image sensors [38–48], to achieve high-quality digital image acquisition and processing. (4) Solar cells: the photoelectric conductivity effect is also one of the working principles of solar cells. By selecting the appropriate photoconductive material, highly efficient and stable solar cells can be made to convert solar energy into electrical energy.

(3) **Photovoltaic effect**

A. **The physical mechanism of the photo photovoltaic effect**

The photovoltaic effect [20] refers to the phenomenon of electromotive force or current generated inside the material when the light shines on the surface of some material. The physical mechanism of this phenomenon mainly involves the interaction of photons and electrons in matter.

Specifically, when a photon (a particle of light) enters the surface of matter, if the energy of the photon is greater than or equal to the electron binding energy in the material, the electron will absorb the photon energy and transition from low to high energy level. In materials with special band structures (such as semiconductors or certain insulators), this transition process may lead to the generation of electron–hole pairs. The electron–hole pair is separated under the action of the internal electric field of the material to form the built-built electric field, thus generating the potential difference at the two ends of the material, that is, the photogenerated volt effect.

The key to this effect is the band structure of the material and the energy of the photon. Materials with proper band structure can absorb photon energy and produce electron–hole pairs, while the photon energy determines the energy level to which electrons can transition and the efficiency of electron–hole pairs.

B. **Classification and characteristics of photovoltaic devices**

Photovoltaic devices are electronic devices made of photovoltaic volt effect, which can be divided into many types according to their structure and characteristics. The following are several common photovoltaic devices and their characteristics:

(a) **Si-based solar cells**: Si-based solar cells are one of the most common photovoltaic devices, with the characteristics of high efficiency, stability, and low cost. It uses monocrystalline Si or polySi as a photoelectric conversion material to convert light energy into electric energy through the photoelectric effect. Si-based solar cells are typically between 15 and 25% efficient and are suitable for large-scale photovoltaic systems.

(b) **Thin film solar cells**: thin film solar cells use thin semiconductor materials (such as amorphous Si, copper indium gallium selenium, etc.) as the photoelectric conversion layer, with the advantages of lightweight, strong flexibility, and low manufacturing costs. Although its efficiency is usually lower than that of Si-based solar cells, thin-film solar cells have broad application prospects in wearable devices, building integration, and other fields.

(c) **Dye-sensitized solar cells**: dye-sensitized solar cells use the principle of dye molecules to absorb photon energy and transfer them to semiconductor materials to achieve photoelectric conversion. This device has the advantages of simple preparation, low cost, and large-area preparation, but its stability and efficiency still need to be further improved.

(d) **Organic solar cells**: organic solar cells use organic materials as the photoelectric conversion layer, with the characteristics of light, flexible, large-area printing and preparation. Although their efficiency is relatively low, organic solar cells have potential applications in wearables, flexible displays, and other fields.

In general, photovoltaic devices have the advantages of high efficiency, environmental protection, and sustainability, and have a wide application prospect in solar power generation, aerospace, building integration, and other fields. As technology advances and costs fall, photovoltaic devices will play a much more important role in the future.

### (6) Photothermoelectric effect

#### A. The principle of the photo thermoelectric effect

Photo thermoelectric effect, also known as the photothermal effect or photothermal conversion effect, is a physical phenomenon that describes the process by which a matter absorbs light energy into heat energy and causes the temperature of the matter to rise. The principle of the photo thermoelectric effect is mainly based on the interaction of photons with electrons or a lattice in matter.

When light hits the surface of the matter, the photons collide with the electron or lattice inside the matter. If the energy of the photon is large enough, it can be absorbed by electrons or lattices in the matter, leading to the vibration of the electron or lattice, which increases the internal energy of the matter, manifested as a rise in temperature. The conversion process of light energy to heat energy is the photo thermoelectric effect.

Different from the photoelectric effect, the photo thermoelectric effect does not directly produce electron–hole pairs or electric currents but affects the temperature of the matter by converting light energy into thermal energy. Therefore, the photo thermoelectric effect is mainly used to measure light intensity and temperature.

#### B. Application of photothermoelectric devices

Photothermoelectric devices are electronic devices made based on the photo thermoelectric effect. They are widely used in various situations where light intensity or temperature change are measured. The following are several common photo thermoelectric devices and their applications:

(a) **Thermistor**: The thermistor is a temperature-sensitive resistor and its resistance value changes with the change of temperature. By measuring the resistance value of the thermistor, the light intensity or the ambient temperature can be measured indirectly. Thermistors are widely used in temperature measurement, light intensity detection, infrared detection, and other fields.

(b) **Thermocouple**: A thermocouple is a device that uses the thermoelectric effect to measure the temperature. It is composed of two wires of different materials. When there is a temperature difference at the junction of the two wires, the electromotive force is generated in the wire. By measuring the size of this electromotive force, the temperature of the junction can be determined. Thermocouples are widely used in temperature measurement, temperature control, and other fields.

(c) **Thermoelectric Reactor**: A thermoelectric reactor is a device composed of multiple thermocouples in series, which can produce a greater electromotive force, thus improving the sensitivity of temperature measurement. Thermoelectric reactors are often used in the fields of high-precision temperature measurement and infrared radiation detection.

The photo thermoelectric effect is an important physical phenomenon with wide applications in optoelectronic detection devices. With their advantages of high sensitivity and high response speed, photo thermoelectric devices play an important role in temperature measurement, light intensity detection, infrared detection, and other fields.

3. **Performance parameters of the photoelectric detection devices**

(1) **Sensitivity**

The sensitivity of a photoelectric detection device refers to its ability to detect and respond to changes in the light signal. Specifically, devices with high sensitivity can detect smaller light signal changes and generate corresponding electrical signals. When measuring the sensitivity of the photoelectric sensor, a standard light source (such as an LED lamp or laser) is usually used to adjust the brightness or distance of the light source, record the intensity of the electrical signal output by the sensor under different conditions, and draw the relationship between the intensity of the light signal and the brightness or distance. By analyzing the curves, the sensitivity range of the sensor as well as the linearity and dynamic range under different optical signal conditions can be determined.

(2) **Response time**

The response time describes the time required from the optical detection device to the generation of a stable output signal. The shorter response time means that the device can respond faster to changes in the optical signal. The response time measurement method includes the square-wave light signal method and the time difference method. The square wave optical signal method determines the response time of the device by measuring the response time of the detector output signal, while the time difference method obtains the response time by measuring the time difference between the output signal and the time-standard signal.

(3) **Noise equivalent power (NEP)**

Noise equivalent power (NEP) is an important indicator of the performance of opto-electronic devices, which describes the minimum optical power that the device can detect. Specifically, when the effective value of the output signal voltage of the photoelectric device is equal to the noise square voltage value, the incident light power is the noise equivalent power. The smaller the value of NEP is, the better the performance of the photoelectric device is, and the fainter the optical signal can be detected.

(4) **Dynamic range**

The dynamic range describes the range of optical signal intensities that the opto-electronic detection device can respond to. It is usually expressed as the ratio of the maximum optical signal intensity to the minimum optical signal intensity, and can also be calculated in units such as f-stop or EV. The wider the dynamic range, the larger the range of optical signal intensity that the device can respond to, which is suitable for a wider range of application scenarios.

(5) **Linear degree**

Linearity is another important performance parameter of the photoelectric detection device, which describes the linear relationship between the device output signal and the measured light signal. Devices with good linearity can more accurately reflect the change of the measured light signal. When measuring linearity, standard light sources of different intensities and the output signal value are recorded. The linearity of the device can be evaluated by comparing the value of the output signal with the signal intensity of the measured light.

5. **Application fields of photoelectric detection devices**

(1) **Photoelectric communication**

Optoelectronic detection devices play a crucial role in the field of optoelectronic communication. Optical communication technology is the main way of modern communication, and the photoelectric detection device is the key component to realize the conversion of optical signals and electrical signals. They are widely used in optical fiber communication, wireless optical communication, and optical carrier communication systems. In fiber optic communication systems, the photoelectric detector is mainly used to receive optical signals transmitted through optical fibers and convert them into electrical signals for further amplification, processing, and transmission. These devices include photodiodes (PDs), and avalanche PDs. Their high sensitivity and high-speed response properties enable long-distance high-speed fiber communication. In the field of wireless optical communication, photoelectric detection devices are equally important. They are used to receive optical signals in the atmosphere or free space and to convert them into electrical signals to enable wireless optical communication. These devices usually require high anti-interference capability and stable performance to cope with complex and changeable environments. Furthermore, in an optical carrier communication system, a photoelectric detection device is used to detect information modulated on the optical carrier and to convert it into an electrical signal for processing. This technology is widely used in cable TV and wireless communications.

(2) **Photoelectric measurement**

Photoelectric detection devices are also widely used in the field of photoelectric measurement. Photoelectric measurement technology is the use of photoelectric effect and photoelectric detection devices to measure a variety of physical quantities, such as light intensity, speed, displacement, Angle, and so on. In light intensity measurement, photoelectric detection devices such as PMTs and PDs are used to detect light intensity, widely used in optical instruments, laser power meters, solar radiometers, and other equipment. In the velocity and displacement measurement, the velocity and displacement of the object can be accurately measured, by detecting the light intensity change when the object passes through the beam. This technology has important applications in production process control, robotics technology, and the automotive industry. In terms of angle measurement, high-precision angle detection can be achieved by using photoelectric detection devices, such as by analyzing photo detectors with beam deflection. This plays an important role in inertial navigation systems, satellite communications, and astronomical observations.

(3) **Optoelectronic control**

In the field of optoelectronic control, optoelectronic detection devices are used to detect and control optical signals and electrical signals. These devices are widely used in automatic control, robotics, optical servo systems, and photoelectric switches. For example, in an automatic control system, the photoelectric detection device can monitor the change of the optical signal in real-time and convert it into electrical signals to adjust the output of the control system and achieve precise control. In robotics, photoelectric detection devices are used to sense the light information in the surrounding environment to guide the action of the robot. In an optical servo system, the photoelectric detection device is used to detect the position or attitude change of the optical element and convert it into electrical signals to drive the corresponding servo mechanism for precise adjustment to ensure the stability and accuracy of the optical system. A photoelectric switch uses a photoelectric detection device to detect the presence or change of optical signal, to realize automatic switch control, widely used in lighting control, safety monitoring and automatic production line and other fields.

(4) **Energy sector**

In the field of energy, photoelectric detection devices also have applications that cannot be ignored. They play an important role in solar cells, photothermal conversion, and energy monitoring. In solar cells, photoelectric detection devices, such as photosensitive materials in solar panels, can convert solar energy into electricity, providing clean and renewable energy for human beings. In the field of photoelectric thermal conversion, photoelectric detection devices can be used to detect and control the light energy and thermal energy conversion efficiency in the process of photoelectric thermal effect, to optimize energy utilization and improve energy conversion efficiency. In terms of energy monitoring, photoelectric detection devices can be used to monitor energy consumption and energy efficiency, such as monitoring and controlling the flow and use of energy in smart grids and energy management systems.

6. **Development trend of photoelectric detection devices**

(1) **R&D and application of new photoelectric materials**

The R&D of new optoelectronic materials is the key to the development of optoelectronic detection devices. Currently, researchers are actively exploring new optoelectronic materials to improve the performance of optoelectronic detection devices. For example, perovskite materials are widely used in solar cells and photoelectric detection fields due to their high absorption coefficient, long carrier life, and easy processing. In addition, 2D materials such as graphene and transition metal sulfides are also used in the development of high-performance optoelectronic detection devices due to their excellent optoelectronic properties [12, 49–55]. The research and development of new optoelectronic materials is not limited to the improvement of the material itself but also includes the composite with other materials to achieve better optoelectronic properties. For example, the combination of perovskite and Si can improve the conversion efficiency of solar cells. At the same time, the research and development of new photoelectric materials also needs to consider their stability and environmental friendliness to achieve the goal of sustainable development.

(2) **The development trends of miniaturization and integration**

With the progress of science and technology, the development trend of miniaturization and integration of photoelectric detection devices is increasingly obvious. Miniaturization reduces device volume reduces cost, and improves its sensitivity and speed. Integration can realize multi-functional integration, and improve the system performance and reliability. The realization of miniaturization and integration relies on advanced manufacturing technology and materials science. For example, microelectronics mechanical system (MEMS) technology enables the miniaturization of optoelectronic detection devices. The integration of photo detector and integrated circuits can realize high-speed and high-sensitivity photoelectric detection.

(3) **Pursuit of high sensitivity, high speed, and high reliability**

The pursuit of high sensitivity, high speed, and high reliability is a key requirement in its practical application. To improve the sensitivity, researchers are exploring new photo absorbing materials and optoelectronic device structures to enhance the collection of photogenerated charges. The realization of high speed requires the development of new optoelectronic detection device structures and readout circuits to reduce the time delay and improve the signal processing speed. In addition, high reliability is an important indicator in the practical application of photoelectric detection devices, and it is necessary to improve their stability and anti-interference ability by optimizing the device structure and material.

(4) **Development trends of intelligence and networking**

With the development of the Internet of Things and big data technology, the trend of intelligence and networking of photoelectric detection devices is becoming more and more obvious. Intelligence can introduce machine learning and artificial intelligence algorithms to realize the adaptive adjustment and self-optimization of photoelectric detection devices and improve their performance and reliability. Networking can connect the photoelectric detection devices with other sensors and network devices to realize real-time data transmission and sharing, to realize remote monitoring and control. This trend will enable optoelectronic detection devices to play an important role in areas such as smart homes, smart transportation, and smart medical care.

## 1.4 Characteristic Parameters of the Photoelectric Detection Device

1. **Overview of the basic characteristic parameters of the photoelectric detection devices**

(1) **Sensitivity**

Sensitivity is a measure of the responsiveness of photoelectric detection devices to changes in the optical signal. It describes the electrical signal intensity output by the device under the unit light signal. The higher the sensitivity, the more sensitive the device is to changes in the optical signal and can detect a weaker optical signal. Sensitivity is usually expressed by the rate of change of voltage or current and is one of the important parameters in the performance evaluation of photoelectric detection devices.

(2) **Response time**

The response time is the time required for the photoelectric detection device to receive the optical signal and generate a stable output signal. It reflects the speed of the device's response to light signal changes. The shorter response time means that the device can respond faster to changes in the optical signal with better dynamic performance. In applications that require rapid response, such as high-speed optoelectronic communication and optical measurements, short response times are important characteristics.

(3) **Dark current**

Dark current refers to the current generated inside the detection device in the absence of light. It is caused by factors such as thermal excitation or material defects. The magnitude of the dark current affects the noise level and sensitivity of the device. A lower dark current means that the device has a lower noise level and can improve the signal-to-noise ratio (SNR) of signal detection.

(4) **Noise equivalent power (NEP)**

Noise equivalent power is an important performance parameter of the photoelectric detection device, which represents the minimum optical signal power that the device can detect. NEP refers to the corresponding incident light signal power when the signal voltage or current output of the device is equal to its internal noise. The smaller the NEP indicates the stronger the ability of the device to detect the weak light signal.

(5) **Dynamic range**

The dynamic range refers to the optical signal power range that the photoelectric detection device can process. It represents the capability of the device to adapt to the detection of both strong and weak light signals. The larger dynamic range means that the device can adapt to a wider range of light signal intensity changes, with better adaptability and flexibility.

(6) **Linearity**

The linearity describes the linear relationship between the output signal and the input optical signal of the photoelectric detection device. Devices with good linearity can reflect the change of light signal more accurately and have higher measurement accuracy and reliability. The linearity is usually expressed by the degree of fit of the calibration curve close to the prescribed line, and its maximum deviation is also known as the nonlinear error.

(7) **Spectral response**

Spectral response refers to the response-ability of a photoelectric detection device to optical signals of different wavelengths. It describes the sensitivity or responsiveness of the device at different wavelengths. Spectral responses are usually obtained by measuring the responsiveness or sensitivity of the device at different wavelengths and plotted as a spectral response curve. The spectral response curves can help us to understand the response characteristics of the device to optical signals at different wavelengths, thus selecting optoelectronic detection devices suitable for specific applications.

The main characteristic parameters of photoelectric detection devices include sensitivity, response time, dark current, noise equivalent power, dynamic range, linearity, and spectral response. These parameters jointly determine the performance and application range of the device, which need to be comprehensively considered according to the specific needs in the selection and application of photoelectric detection devices.

2. **Effect of characteristic parameters on the performance of photoelectric detection devices**

(1) **The influence of sensitivity on performance**

Sensitivity is one of the important indicators of the performance of photoelectric detection devices, which characterize the response of the device to optical signals. The higher the sensitivity, the stronger the device can detect the optical signal and can detect a weaker optical signal. For photoelectric detection devices, improving the sensitivity can increase their application range in communication, detection, and other fields. However, improved sensitivity often requires the sacrifices of other performance, such as noise-equivalent power.

(2) **The influence of response time on performance**

The response time is the time required for the photoelectric detection device to receive the optical signal and stabilize the output signal. The shorter the response time is, the faster the response speed of the device to the optical signal, which is crucial for application scenarios with high real-time requirements. However, a shortened response time may lead to a reduction in sensitivity due to the need to reduce device size and increase capacitance, etc. Therefore, a trade-off between response time and sensitivity is required when designing optoelectronic detection devices.

(3) **The influence of the noise-equivalent power on the performance**

Noise equivalent power refers to the noise power generated by a photoelectric detection device within a certain bandwidth, which has an important influence on the performance of the device. The lower the noise equivalent power, the higher the SNR of the device and the better the quality of the detected signal. However, reducing the noise-equivalent power often requires increasing the device size and power consumption, which may affect the response speed and sensitivity of the device. Therefore, comprehensive considerations between noise-equivalent power, response time, and sensitivity are required when designing and manufacturing optoelectronic detection devices.

(4) **The influence of the dynamic range on the performance**

Dynamic range refers to the signal intensity range that the photoelectric detection device can process. A larger dynamic range indicates that the device can adapt to a stronger optical signal without overload or a low SNR. It is of great significance to improve the dynamic range for the application scenarios with high dynamic range requirements, such as high-speed communication and strong light background. However, increasing the dynamic range may lead to reductions in sensitivity and response time, thus requiring trade-offs based on demand in practice.

(5) **The influence of linearity on performance**

Linearity refers to the linear relationship between the output signal and the input optical signal of the photoelectric detection device. The better the linearity is, the closer the relationship between the output signal and the input signal is, and the more accurately it can reflect the change of the optical signal. Improving the linearity is beneficial to improve the application performance of optoelectronic detection devices in fields such as simulation and digital communication. However, the improvement of the linearity often requires increasing the device size and complexity, which may affect the device response speed and sensitivity. Therefore, a trade-off between linearity, response time, and sensitivity is required in the design and fabrication of optoelectronic detection devices.

The basic characteristic parameters of optoelectronic detection devices have important effects on their performance and application range. When selecting and applying optoelectronic detection devices, it is necessary to evaluate their characteristic parameters according to the specific application requirements and requirements and select the most suitable optoelectronic detection devices.

3. **Measurement and optimization of characteristic parameters**

(1) **Measurement method of characteristic parameters**

The characteristic parameter is a key index to evaluate the performance of photoelectric detection devices, mainly including responsiveness, noise equivalent power, bandwidth, quantum efficiency, half-wave width, etc. Measuring these characteristic parameters usually requires professional testing equipment and methods.

A. **Measurement of responsiveness**

Responsiveness is the degree to which a photoelectric detection device responds to the optical signal, usually expressed in R in A/W. The responsiveness is measured mainly by measuring the relationship between the current or photocurrent of the device at a specific light intensity and the incident light power. In the experiment, the intensity of the incident light is gradually changed, and the change of the detector output current is monitored to draw the I-L curve (photocurrent-light intensity curve). The slope of the curve is the responsiveness, and the greater the responsiveness, the stronger the response of the device to the optical signal.

B. **Measurement of noise-equivalent power**

Noise equivalent power (NEP) refers to the thermal noise power equal to the noise equivalent power generated by the ideal noise-free detector at the same bandwidth and temperature at a certain SNR condition. Measuring NEP usually requires a noise analyzer, calculated by the measurement of the noise power spectrum under specific bandwidth and temperature conditions. The smaller the NEP indicates the better the noise performance of the detector.

C. **Bandwidth measurement**

Bandwidth is an important parameter to describe the response speed of photoelectric detection devices, which determines the highest frequency of optical signals that the device can process. The method of measuring the bandwidth is usually to conduct a fast Fourier transform (FFT) analysis of the detector output signal to determine the frequency corresponding to the main part of the signal energy. Furthermore, the bandwidth can also be estimated by observing the rise time or downtime of the detector response curve.

D. **Measurement of quantum efficiency**

Quantum efficiency ($\eta$) refers to the ratio of electron–hole pairs generated by incident photons, reflecting the efficiency of the device in converting light energy into electric energy. Measuring quantum efficiency usually requires comparing the output current of the detector at known light intensity with the theoretical output current in the ideal situation. In the experiment, the photocurrent at different light intensities is measured and the quantum efficiency is calculated. The higher the quantum efficiency is, the higher the photoelectric conversion efficiency of the device.

E. **Measuring Full-Width-Half-Maximum**

Full-Width-Half-Maximum (FWHM) refers to the half-peak width of the detector response curve, which is a parameter describing the response characteristics of the device. Measuring the half-wave width usually requires drawing I-L curves and then determining by the half-peak point on the curve.

(2) **Methods to improve the characteristic parameters**

A. **Optimize the device structure**

To improve the performance of the photoelectric detection devices, it can be achieved by optimizing the device structure. For example, a multilayer structure design can be adapted to enable the photogenerated carriers to be separated even more efficiently inside the device, thus improving the response speed and quantum efficiency of the photoelectric detection device. Furthermore, the parasitic capacitance can be reduced by reducing the device size to further improve the speed and sensitivity of the device.

B. **Material selection**

The choice of materials has important effects on the performance of optoelectronic detection devices. For example, a material with a high absorption coefficient and good electrical characteristics can be selected as the optical absorption layer to improve the photoelectric conversion efficiency of the device. In addition, materials with high mobility and low noise can be selected as the transmission layer to improve the response speed and SNR of the device.

C. **Process optimization**

In the fabrication of optoelectronic detection devices, process optimization is crucial to improve device performance. For example, the thickness and crystallization quality of the optical absorption layer can be optimized by controlling process parameters such as sputtering power and annealing temperature. In addition, the surface defects and surface state density of the device can be reduced.

D. **Circuit design**

Circuit design is also one of the important factors affecting the performance of photoelectric detection devices. The SNR and stability of the device can be improved by optimizing the bandwidth and gain of the preamplifier, as well as by using suitable filters and bias circuits. In addition, the noise and performance of the device can be further reduced by using a differential amplifier and low noise amplifier.

(3) **The influence of the optimization of the characteristic parameters on the device performance**

The optimization of the characteristic parameters directly affects the performance of the optoelectronic detection devices. For example, improving the responsiveness can increase the photoelectric conversion efficiency of the device; increasing the bandwidth can broaden the application range of the device; improving the quantum efficiency can increase the SNR of the device; and reducing the noise equivalent power can improve the detection sensitivity of the device. Therefore, the performance of the optoelectronic detection devices can be significantly improved through the optimization of the characteristic parameters.

# References

1. Sze S (1981) Physics of semiconductor devices. John Wiley & Sons, New York
2. Ellmer K (2012) Hall effect and conductivity measurements in semiconductor crystals and thin films. In: Characterization of materials, pp 1–16
3. Look D (2002) Hall effect in semiconductors. In: Characterization of materials
4. Gaj JA (2011) 2.04—Semimagnetic semiconductors. In: Bhattacharya P, Fornari R, Kamimura H (eds) Comprehensive semiconductor science and technology. Elsevier, Amsterdam, pp 95–124
5. Cova S, Ghioni M, Lacaita A, Samori C, Zappa F (1996) Avalanche photodiodes and quenching circuits for single-photon detection. Appl Opt 35(12):1956. https://doi.org/10.1364/ao.35.001956
6. Liu J, Xia F, Xiao D, García de Abajo FJ, Sun D (2020) Semimetals for high-performance photodetection. Nat Mater 19(8):830–837. https://doi.org/10.1038/s41563-020-0715-7
7. Dilello NA, Hoyt JL (2011) Impact of post-metallization annealing on Ge-on-Si photodiodes passivated with silicon dioxide. Appl Phys Lett 99(3):10–13. https://doi.org/10.1063/1.3615800
8. Lopez-Sanchez O, Dumcenco D, Charbon E, Kis A (2014) Avalanche photodiodes based on $MoS_2$/Si heterojunctions
9. Liu Y, Duan X, Shin HJ, Park S, Huang Y, Duan X (2021) Promises and prospects of two-dimensional transistors. Nature 591(7848):43–53. https://doi.org/10.1038/s41586-021-03339-z
10. Baugher BWH, Churchill HOH, Yang Y, Jarillo-Herrero P (2014) Optoelectronic devices based on electrically tunable p-n diodes in a monolayer dichalcogenide. Nat Nanotechnol 9(4):262–267. https://doi.org/10.1038/nnano.2014.25
11. Fu W, Jiang L, van Geest EP, Lima LMC, Schneider GF (2017) Sensing at the surface of graphene field-effect transistors. Adv Mater 29(6):1603610. https://doi.org/10.1002/adma.201603610
12. Li L et al (2014) Black phosphorus field-effect transistors. Nat Nanotechnol 9(5):372–377. https://doi.org/10.1038/nnano.2014.35
13. Cotter JE, Guo JH, Cousins PJ, Abbott MD, Chen FW, Fisher KC (2006) P-type versus n-type silicon wafers: prospects for high-efficiency commercial silicon solar cells. IEEE Trans Electron Devices 53(8):1893–1901. https://doi.org/10.1109/TED.2006.878026
14. Pospischil A, Furchi MM, Mueller T (2014) Solar-energy conversion and light emission in an atomic monolayer p-n diode. Nat Nanotechnol 9(4):257–261. https://doi.org/10.1038/nnano.2014.14
15. Allen TG, Bullock J, Yang X, Javey A, De Wolf S (2019) Passivating contacts for crystalline silicon solar cells. Nat Energy 4(11):914–928. https://doi.org/10.1038/s41560-019-0463-6
16. Hu Q (2024) Advancements and prospects in third-generation semiconductor materials: a comprehensive analysis. Highlights Sci Eng Technol 81:631–636. https://doi.org/10.54097/cbcyx445
17. Zhang Y (2020) The application of third generation semiconductor in power industry. E3S Web Conf 198:0–5. https://doi.org/10.1051/e3sconf/202019804011
18. Yu Q (2024) Comparative analysis of Sic and Gan: third-generation semiconductor materials. Highlights Sci Eng Technol 81:484–490. https://doi.org/10.54097/2q3qyj85
19. Lou LF, Kitazaki KS (1989) Photon emission from metal-oxide-semiconductor capacitors under fast-ramp conditions. J Appl Phys 66(11):5618–5621. https://doi.org/10.1063/1.343668
20. Priolo F, Gregorkiewicz T, Galli M, Krauss TF (2014) Silicon nanostructures for photonics and photovoltaics. Nat Nanotechnol 9(1):19–32. https://doi.org/10.1038/nnano.2013.271
21. Pool PJ, Morris DG, Burt DJ, Bell RT, Holland AD, Smith DR (2005) Application of electron multiplying CCD technology in space instrumentation. In: Proceedings SPIE vol 5902, 59020A. https://doi.org/10.1117/12.621627

22. Gendreau KC, Prigozhia GY, Huang RK, Bautz MW (1995) A technique to measure trap characteristics in CCD's using x-rays. IEEE Trans Electron Devices 42(11):1912–1917. https://doi.org/10.1109/16.469396
23. Matsuura D, Nakajima H, Miyata E, Tsunemi H, Doty JP, Ikeda H (2007) Development of an ASIC for multi-readout X-ray CCDs. IEEE Nucl Sci Symp Conf Rec 2:1073–1077. https://doi.org/10.1109/NSSMIC.2007.4437195
24. IEEE ED 2003 EMCCD. The noise performance of electron multiplying CCD.pdf
25. Molecular_expressions. Concepts in digital imaging technology: CCD noise sources and signal-to-noise ratio. https://andor.oxinst.com/learning/view/article/ccd-signal-to-noise-ratio
26. Smith GE (2010) Nobel lecture: the invention and early history of the CCD. Rev Mod Phys 82(3):2307–2312. https://doi.org/10.1103/RevModPhys.82.2307
27. Smith GE (2011) The invention and early history of the CCD. J Appl Phys 109(10). https://doi.org/10.1063/1.3578638
28. Dussault D, Hoess P (2004) Noise performance comparison of ICCD with CCD and EMCCD cameras. In: Infrared systems and photoelectronic technology, vol 5563, October, p 195. https://doi.org/10.1117/12.561839
29. Ali M et al (2022) In-situ monitoring of reciprocal charge transfer and losses in graphene-silicon CCD pixels. Sensors 22(23):9341. https://doi.org/10.3390/s22239341
30. Yang S, Guo M, Li B, Xia J, Wang Q, Sun F (2012) Design of high resolution high sensitivity EMCCD camera. In: 2012 Second international conference on intelligent system design and engineering application, pp 1212–1216. https://doi.org/10.1109/ISdea.2012.582
31. Touron P, Roy F, Magnan P, Marcelot O, Demiguel S, Virmontois C (2020) Capacitive trench-based charge transfer device. IEEE Electron Device Lett 41(9):1388–1391. https://doi.org/10.1109/LED.2020.3014431
32. Zhou J et al (2022) Charge sampling photodetector based on van der Waals heterostructures. Adv. Opt. Mater. 10(24):1–8. https://doi.org/10.1002/adom.202201442
33. Borg J, Johansson J (2011) Evaluation of a surface-channel CCD manufactured in a pinned active-pixel-sensor CMOS process. IEEE Trans Electron Devices 58(8):2660–2664. https://doi.org/10.1109/TED.2011.2156798
34. Clarke AS, Hall DJ, Holland A, Burt D (2012) Modelling charge storage in Euclid CCD structures. J Instrum 7(1). https://doi.org/10.1088/1748-0221/7/01/C01058
35. Burke BE, Member S, Mountain RW, Daniels PJ, Cooper MJ, Dolat VS (1994) CCD soft x-ray imaging spectrometer for the ASCA satellite. 41(1)
36. White MH, Blaha FC, Lampe DR, Mack IA (1974) Characterization of surface channel CCD image arrays at low light levels. IEEE J Solid-State Circuits 9(1):1–12. https://doi.org/10.1109/JSSC.1974.1050448
37. Leitz C et al (2017) Germanium CCDs for large-format SWIR and X-ray imaging. J Instrum 12(5). https://doi.org/10.1088/1748-0221/12/05/C05014
38. Luštica A (2011) CCD and CMOS image sensors in new HD cameras. In: Proceedings Elmar—international symposium electronics in marine, September, pp 133–136
39. El Gamal A, Eltoukhy H (2005) CMOS image sensors. IEEE Circuits Devices Mag 21(3):6–20. https://doi.org/10.1109/MCD.2005.1438751
40. Fossum ER, Member S (1997) CMOS image sensors: electronic camera-on-a-chip. 44(10):1689–1698
41. Kim WT, Park C, Lee H, Lee I, Lee BG (2019) A high fullwell capacity CMOS image sensor for space applications. Sensors (Switzerland) 19(7):1505. https://doi.org/10.3390/s19071505
42. Shimizu R et al (2009) A charge-multiplication CMOS image sensor suitable for low-light-level imaging. IEEE J Solid-State Circuits 44(12):3603–3608. https://doi.org/10.1109/JSSC.2009.2035541
43. Bigas M, Cabruja E, Forest J, Salvi J (2006) Review of CMOS image sensors. Microelectronics J 37(5):433–451. https://doi.org/10.1016/j.mejo.2005.07.002
44. Goossens S et al (2017) Broadband image sensor array based on graphene-CMOS integration. Nat Photonics 11(6):366–371. https://doi.org/10.1038/nphoton.2017.75

45. Fossum ER et al (1998) CMOS active pixel image sensors fabricated. IEEE Trans Electron Devices 45(4):889–894
46. Fossum ER (1997) CMOS image sensors: electronic camera-on-a-chip. IEEE Trans Electron Devices 44(10):1689–1698. https://doi.org/10.1109/16.628824
47. Malik AI, Grimalsky V (2004) CMOS infrared sensors with giant internal signal amplification. In: Ultrahigh-and high-speed photography, photonics, and videography, vol 5210, p 124. https://doi.org/10.1117/12.508019
48. Shimizu R, Arimoto M, Nakashima H, Misawa K, Ohno T, Nose Y (2009) A charge-multiplication CMOS image sensor suitable for low-light-level imaging. 44(12):3603–3608
49. Seo J et al (2021) Ultrasensitive photodetection In $MoS_2$ Avalanche phototransistors. Adv Sci 8(19):1–9. https://doi.org/10.1002/advs.202102437
50. Wang C et al (2022) Neuromorphic device based on silicon nanosheets. Nat Commun 13(1):1–10. https://doi.org/10.1038/s41467-022-32884-y
51. Das S (2016) Two dimensional electrostrictive field effect transistor (2D-EFET): a sub-60 mV/decade steep slope device with high on current. Sci Rep 6(July):1–7. https://doi.org/10.1038/srep34811
52. Ates HC et al (2022) End-to-end design of wearable sensors. Nat Rev Mater 7(11):887–907. https://doi.org/10.1038/s41578-022-00460-x
53. Novoselov KS, Mishchenko A, Carvalho A, Castro Neto AH (2016) 2D materials and van der Waals heterostructures. Science (80-) 353(6298). https://doi.org/10.1126/science.aac9439
54. Amani M, Regan E, Bullock J, Ahn GH, Javey A (2017) Mid-wave infrared photoconductors based on black phosphorus-arsenic alloys. ACS Nano 11(11):11724–11731. https://doi.org/10.1021/acsnano.7b07028
55. Furchi MM, Polyushkin DK, Pospischil A, Mueller T (2014) Mechanisms of photoconductivity in atomically thin $MoS_2$. Nano Lett 14(11):6165–6170. https://doi.org/10.1021/nl502339q

# Chapter 2
# Common Light Source in Photoelectric Detection

**Abstract** This chapter explores the characteristics and applications of light sources in photoelectric detection systems, focusing on their impact on performance, selection, and optimization. It categorizes light sources into natural and artificial types, detailing their roles in detecting temperature, pressure, and velocity. Key parameters (i.e. brightness, spectral distribution, stability, lifespan) and various light sources (i.e. thermal radiation sources, gas discharge lamps) are discussed, alongside applications in biomedical imaging, security monitoring, and communication systems. The chapter discusses spectral properties and emission mechanisms to provide insights into radiation energy distribution, and also explores advancements in light source technology. It concludes by highlighting the evolution of LED and laser technologies, particularly in fluorescence detection, medical diagnostics, and sensor calibration.

**Keywords** Light source · Bandwidth · Central wavelength · Radiation laws · Laser · Light-emitting diode · Gas discharge · Stability · Thermal radiation sources

## 2.1 Basic Characteristic Parameters of the Light Source

1. **Overview of the light source**

(1) **Definition and classification of the light source**

A light source is a physical term referring to an object capable of emitting electromagnetic waves over a certain range of wavelengths, including both visible light and invisible light, such as ultraviolet, infrared, and X-rays. In our daily lives, we are most often concerned with objects that emit visible light, such as the sun, electric lights, and burning candles.

Based on the method of generation, light sources can be categorized into two main types: natural light sources and artificial light sources. Natural light sources exist naturally, such as the sun, stars, and bioluminescent organisms like fireflies and lanternfish. Artificial light sources, on the other hand, are created through human ingenuity and include torches, oil lamps, candles, incandescent lamps, fluorescent lamps, LEDs, and more.

 
S. Du and W. Yin, *Photodetection and Image Sensing Techniques*, Advanced Topics in Science and Technology in China 2, https://doi.org/10.1007/978-981-96-5479-6_2

(2) **Function of the light source in the photoelectric detection system**

In a photoelectric detection system, the light source is an essential component. The beam of light it emits interacts with the object being tested, serving as a carrier of the information to be measured. The detected object may involve various physical quantities, such as temperature, pressure, or velocity. When the light beam strikes the object, it undergoes conversion into an optical signal carrying the characteristic information of the measured subject through various optical effects, such as interference, reflection, or absorption. This optical signal is then received by a photo detector and converted into an electrical signal for subsequent signal processing and analysis.

The selection of the light source in photoelectric detection systems is critical. The choice must align with the specific requirements of the application, including considerations of radiation power, spectral range, and luminous spatial distribution. In some cases, the light source itself may be the object of measurement. For instance, in certain biological detection or environmental monitoring applications, it may be necessary to measure characteristics of the light source itself, such as its spectral distribution or light intensity.

(3) **The influence of the light source on the photoelectric detection performance**

The influence of the light source on the photoelectric detection performance is mainly reflected in the following aspects:

A. **The brightness of the light source**

The brightness of a light source directly impacts the performance of photoelectric detection. Higher brightness increases the photon energy, which in turn enhances detection efficiency. However, excessively high brightness can overshine the photo detector, leading to reduced performance. Therefore, in practical applications, it is crucial to select a light source with suitable brightness, tailored to the characteristics and requirements of the photo detector.

B. **Spectral distribution of the light source**

The spectral distribution of the light source is also crucial for photoelectric detection performance. Different photo detectors exhibit varying levels of responsiveness to light at different wavelengths. Therefore, the spectral distribution of the light source must align with the response wavelength of the photo detector to achieve optimal detection performance. Furthermore, the spectral distribution can influence the brightness and stability of the light source, requiring a comprehensive evaluation during selection.

C. **Stability of the light source**

The stability of the light source is also very important to affect the photoelectric detection performance. The stability of the light source includes the brightness stability and spectral stability of the light source. If the stability of the light source is not good, the output signal of the photo detector will fluctuate, which affects its performance. Therefore, it is necessary to choose a stable light source or take corresponding measures to reduce the influence of light source fluctuations on the performance of the photo detector.

D. **Life span of the light source**

The lifetime of the light source also affects the photoelectric detection performance. If the light source has a short life and needs frequent replacement, it will not only increase the cost but also affect the stability and reliability of the photoelectric detection system. Therefore, when choosing a light source, we need to consider its life, and how to extend its service life.

The light source plays a crucial role in the photoelectric detection system. The performance and application range of the photoelectric detection system can be improved by rationally selecting and using light sources.

2. **Basic characteristics and parameters of the light source**

(1) **Spectral characteristics**

A. **Spectral distribution**

The spectral distribution is a parameter describing the intensity distribution of the light emitted by the light source in the wavelength direction. Different light sources have different spectral distributions, which directly reflect the type and quality of the light source. For example, the spectral distribution of sunlight shows a continuous distribution feature, while the spectral distribution of laser light is usually very narrow and concentrated at a specific wavelength. The measurement of the spectral distribution is usually conducted by a spectrometer, and the radiation characteristics of the light source can be obtained by analyzing the spectral distribution, such as the peak wavelength, the spectral half-width, etc.

B. **Central wavelength**

The central wavelength refers to the wavelength at which the light emitted by a source is most intense within its spectrum. This parameter is crucial for various applications. For instance, in fiber communication, the laser's wavelength must match the transmission window of the optical fiber, requiring precise control of the laser's central wavelength. The central wavelength is usually measured with a spectrometer, which determines it by locating the strongest peak in the spectrum.

C. **Spectral bandwidth**

Spectral bandwidth refers to the wavelength range where the light intensity decreases to $1/e$ of its maximum value (approximately 36.8%) in the spectrum of the light emitted by the light source. This parameter indicates the spectral purity of the light source; the smaller the bandwidth, the purer the spectrum. To measure the spectral bandwidth, a spectrometer is typically used. The measurement is determined by calculating the wavelength range where the light intensity, near the peak wavelength, drops to $1/e$ of the maximum value.

(2) **Optical power and light intensity**

A. **Light power**

Optical power refers to the light energy emitted by a light source per unit of time, which is usually expressed by a unit of "watt" (W). It is an important parameter to measure the brightness of the light source. The size of the optical power depends on the type of light source, the luminous efficiency, and the operating voltage. In practical application, optical power has a close relationship with optical fiber communication, lighting, display technology, and other fields. For example, in optical fiber communication, the optical power at the transmitting end needs to reach a certain value to ensure that the optical signal is not lost due to attenuation during the transmission in the optical fiber.

B. **Light Intensity**

Light intensity refers to the power of light emitted by a light source in a specific direction per unit solid angle, with the unit "candela" (cd). It is a key parameter for describing the luminous intensity of a light source, and is closely related to the brightness, luminous area, and luminous angle of the source. The distribution of light intensity plays a crucial role in technologies such as lighting and display systems. For example, in lighting design, the range and intensity of lighting should be determined based on the light intensity distribution characteristics of the source to achieve the desired lighting effect.

(3) **Stability of the light source**

A. **Temporal stability**

Temporal stability refers to the degree of change in the light performance (such as light power, light intensity, etc.) of the light source during a long time of operation. The good time stability of the light source means that it can maintain a stable luminescence performance during the long time of use, and will not significantly reduce the luminescence efficiency due to the passage of time. Temporal stability has an important influence on the service life and reliability of the light source. In the design and selection of the light source, its temporal stability needs to be considered to ensure that the light source can meet the long-term and stable luminescence requirements in practical application.

B. **Temperature stability**

Temperature stability refers to the degree to which the light performance of a light source at different ambient temperatures. The good temperature stability of the light source means that it can maintain a stable luminescence performance under different temperature conditions, and will not change the light power and light intensity significantly due to the temperature change. Temperature stability has an important influence on the applicability of the light source in various environments. In practical applications, light sources may encounter high or low-temperature environments, so good temperature stability is required to ensure that it works properly at different temperatures.

(4) **Life span of the light source**

The life of the light source refers to the time when the light emitting can be kept within the specified range during the operation of the light source, and the common unit is hours. The life of different light sources is very different. For example, the life of light sources such as fluorescent light and LED light is generally more than 10,000 h, while the life of incandescent light sources is only 1000–2000 h. The correct estimation of the life span of the light source is very important for both investors and users. The longer the light source life, the shorter the service life, the lower the service cost; and the short light source life means long service life and high service cost.

(5) **The divergence angle of the light source**

The light source divergence angle refers to the angle at which the light beam spreads after being emitted from the source over a certain distance. This angle reflects the extent of beam diffusion. A smaller divergence angle results in a higher energy density, a smaller irradiation area, and stronger light intensity. Conversely, a larger divergence angle leads to a lower energy density, a larger irradiation area, and weaker light intensity. Therefore, the magnitude of the beam's divergence angle plays a crucial role in the irradiation effect of the beam. In practical applications, it is essential to select the appropriate light source and adjust the beam divergence angle based on the specific scene and irradiation requirements.

(6) **Modulation characteristics of the light source**

Light source modulation refers to the process of altering the output parameters of a light source to transmit desired information. This allows for the transmission of information by adjusting certain properties of the light source, such as light intensity, frequency, or phase. Internal modulation involves directly applying the modulation signal to the light source, causing it to modulate its light parameters. This method is also known as direct modulation. For example, a semiconductor laser or a light-emitting diode (LED) can use direct modulation. In the case of a semiconductor laser, the modulation signal, along with the bias current, must exceed the threshold for modulation to occur. In contrast, external modulation involves a light modulator located outside the light source. The light modulator independently controls the light parameters, providing greater flexibility for modulating the light source.

3. **Measurement and evaluation of light source characteristic parameters**

(1) **Measurements of spectral characteristics**

Measurements of the spectral properties are usually performed using either a spectral analyzer or a spectrometer. These devices can measure the wavelength and intensity distribution of the light emitted by the source, thus obtaining the spectral properties of the source. In the measurement, the light from the light source should be focused through the optical system and illuminated on the incident slit of the spectrometer. Then the spectrometer will disperse the light according to the wavelength of the light, measure the light intensity on each wavelength through the detector, and finally get the spectral characteristics of the light source.

(2) **The measurements of light power and light intensity**

Measurement of the optical power and intensity can be performed using an optical power meter and a photometer. The optical power meter can directly measure the power of the light emitted by the light source, while the photometer can measure the intensity of the light emitted by the light source. During the measurement, it is necessary to point the probe of the optical power meter or photometer at the light emitted by the light source, and then the measurement result is read through the display screen or software on the device.

(3) **Evaluation method of light source stability**

Assessment of light source stability can be achieved by monitoring the power or intensity of the light emitted by the light source for a long time. During the evaluation process, the optical power meter or photometer can be used to continuously measure the power or light intensity of the light emitted by the light source and record the measurement results. The stability of the light source can be assessed by analyzing the variation of the measurements. If the power or intensity of the light emitted by the light source remains relatively stable for a long time, it indicates that the light source has good stability.

(4) **Prediction and testing of light source life**

The prediction of the life of the light source is usually based on factors such as the manufacturing material of the light source, working conditions, and the use environment of the light source. During the test process, the performance change of the light source can be observed by simulating the use of the light source under the actual working conditions, such as continuous work, temperature change, humidity change, etc. By recording the working life of the light source under different conditions, we can predict the life of the light source in actual use. In addition, the accelerated aging experiment and other methods can also be used to accelerate the aging process of the light source, shorten the test time, and obtain more accurate life prediction results.

(5) **Measurement of the divergence angle**

The measurement of the divergence angle of the light source can be performed using an integral sphere, beam analyzer, and other equipment. The integral sphere can capture all the light entering the sphere and calculate the divergence angle of the light source by measuring the light intensity at different positions within the sphere. The light beam analyzer can directly measure the divergence Angle of the light beam emitted by the light source. During measurement, it is necessary to place the light source in a specific position of the integral sphere or beam analyzer and adjust the equipment to measure the divergence angle of the beam emitted by the light source.

(6) **Test method of light source modulation characteristics**

The test of light source modulation characteristics can be conducted using an oscilloscope, signal generator, and other equipment. During the test, a signal generator is used to generate a modulation signal of a certain frequency and amplitude and apply it to the light source. Then, an oscilloscope is used to measure the modulation signal of the light emitted by the light source, and the waveforms and parameters of the modulation signal are observed. The modulation properties of the light source can be evaluated by analyzing the waveform and parameters of the modulation signal. If the waveform of the modulation signal is clear and the parameters are stable, it indicates that the light source has better modulation characteristics.

4. **Application of light source characteristic parameters in photoelectric detection**

(1) **The influence of the light source characteristic parameters on the performance of the photoelectric detection system**

A. **Spectral properties**

Spectral characteristics are one of the most important factors in the characteristic parameters of light sources on the performance of photoelectric detection systems. The spectral properties of the light source directly determine the response range and sensitivity of the photodetection system to the optical signal. For a specific wavelength of light, the responsiveness of the photoelectric detection system will be maximized, thus improving the performance of the detection system. Therefore, the spectral characteristics are needed to ensure the best performance of the photoelectric detection system.

B. **Light power and light intensity**

Optical power refers to the light energy emitted by a light source per unit of time, which is usually expressed in watts (W). Light intensity refers to the light power emitted by the light source in the unit stereo Angle, usually expressed by the unit per sphere per watt (W/sr). The optical power and intensity of the light source have a direct impact on the performance of the photoelectric detection system. The higher the optical power, the higher the SNR ratio of the detection system, and the better the detection performance. The light intensity determines the radiation range of the light source required to be received by the photoelectric detection system. The greater the light intensity, the farther the detection distance and range of the detection system are.

C. **Light source stability**

The stability of the light source refers to the degree of change of the light power, spectral distribution, and other characteristic parameters during the long-time operation. Source stability has an important influence on the performance of photoelectric detection systems. If the light source is stable, it will fluctuate in the performance of the detection system and affect the accuracy of the detection results. Therefore, in the practical application, it is necessary to select a light source with good stability, and the real-time monitoring and maintenance of the light source, to ensure the stability of the photoelectric detection system.

D. **Light source life**

Light source life refers to the time when the light source is used until it does not work properly. The life of the light source has a direct impact on the service cost and maintenance cycle of the photoelectric detection system. In general, the longer the life of the light source, the lower the use cost of the detection system, and the longer the maintenance cycle. However, the life of the light source and the spectral characteristics, optical power, and other factors have some contradictions, so in the design and selection of the light source, it is necessary to consider all factors to determine the most suitable light source life.

E. **Divergence angle of the light source**

The divergence Angle of the light source refers to the diffusion degree of the light emitted by the light source in the space. The divergence angle of the light source has an important influence on the detection range and accuracy of the photoelectric detection system. In general, the smaller the divergence angle of the light source, the smaller the detection range of the detection system, but the higher the accuracy. However, too small the divergence angle of the light source will lead to light energy concentration, which may affect the service life of the light source. Therefore, in practical applications, we need to choose the appropriate divergence angle according to the specific detection requirements and environmental conditions.

F. **Light source modulation characteristics**

The modulation characteristic of the light source refers to the change law of the light intensity under the action of the modulation signal. The modulation characteristics of the light source have a direct impact on the modulation performance and signal processing ability of the photoelectric detection system. In general, the better the modulation characteristics of the light source, the stronger the modulation performance and signal processing capability of the detection system. Therefore, the modulation characteristics of the light source are needed to ensure the optimal performance of the photoelectric detection system.

(2) **Select the appropriate light sources according to the detection requirements**

When selecting the light source of the photoelectric detection system, it is necessary to choose the appropriate light source type and parameters according to the detection target and requirements. For example, in biomedical imaging, it is necessary to select light sources with specific spectral distribution and high optical power to penetrate biological tissues and obtain clear images; in security monitoring, it is necessary to select light sources with high stability and long life to ensure the stable operation of the system for a long time; in a communication system, it is necessary to choose light sources with good modulation characteristics to realize high-speed data transmission.

(3) **Optimization and adjustment of the light source characteristic parameters to improve the detection performance**

In the photoelectric detection system, the detection performance can be improved by optimizing and adjusting the light source characteristic parameters. For example, the spectral absorption characteristics of the light source can be matched by adjusting the spectral distribution; the detection distance and sensitivity can be increased by increasing the optical power and intensity of the light source; the reduction of the signal fluctuation and noise; the maintenance cost of the light source can be reduced by increasing the life of the light source; and the reduction of the beam of the light source by optimizing the modulation characteristics of the light source. These optimization and adjustment measures can be selected and implemented according to the specific detection requirements and objectives.

## 2.2 Thermal Radiation Source

1. **Overview of the thermal radiation sources**

A thermal radiation source is an important concept involved in thermodynamics and optics. The following is an overview of it:

(1) **Define the thermal radiation source**

A heat radiation source, also known as a heat light source, is a light source that can emit visible light to the surroundings, but also emit thermal radiation to the surrounding air. Its main energy comes from high-temperature objects, and compared to the phenomenon of electromagnetic waves emitted through optical devices or artificial sources, thermal light sources are common in nature, such as the sun, flame, and so on. The formation mechanism of thermal radiation light source originates from the principle of thermodynamics, that is, the temperature of an object has a direct relationship with its radiation energy. The higher the temperature, the stronger the radiation energy of the object, and the brighter the light obtained.

(2) **Classification of thermal radiation sources**

Heat radiation sources can be classified according to their generation mode and characteristics. A common classification is based on the temperature of the radiation source and the type of radiation. For example, planar thermal radiation is radiation from the surface of objects with a certain temperature, such as thermal radiation from the surface of the sun and the surface of human skin. Heat disease radiation is the radiation caused by the constant vibration and collision of molecules in high-temperature objects. Typical examples include the radiation from flame and molten metal. In addition, there is thermal burst radiation, which is due to the rapid movement of high-energy particles, usually occurring in extremely high-energy events such as nuclear explosions, supernova bursts, and lasers.

(3) **The role and status of thermal radiation sources in photoelectric detection**

Thermal radiation sources play an important role in photoelectric detection systems. First of all, the thermal radiation source is an important signal source in the photoelectric detection system, and the thermal radiation emitted by it can be received by the photo detector and converted into an electrical signal, and then realize the detection, identification, or measurement of the measured object. Secondly, the characteristics of the thermal radiation source have important effects on the performance of the photoelectric detection system. For example, the spectral characteristics, optical power, stability, and other parameters of the thermal radiation source will affect the detection distance, sensitivity, SNR, and other performance indicators of the photoelectric detection system. Therefore, in the design and selection of a photoelectric detection system, it is necessary to choose the appropriate heat radiation source according to the actual requirements and optimize and adjust accordingly.

As an important part of the photoelectric detection system, the performance and application of the thermal radiation source directly affect the development and application of the photoelectric detection technology. With the continuous progress of science and technology and the increasing demand for application demand, the research and application of heat radiation sources will be more widely paid attention and attention.

2. **Basic characteristics of thermal radiation sources**

(1) **Blackbody radiation laws**

A. **Planck's Law of Radiation**

The central idea of Planck's law of radiation is to quantize the radiation energy. According to Planck's theory, the radiation energy exists in a discrete energy quantum form, and the size of the energy quantum is proportional to the radiation frequency, namely $E = hf$, where E is the energy, h is the Planck constant and f is the frequency. The law has extensive applications in physics and astronomy, revealing the dependence of the radiation energy density with wavelength and the dependence of the peak position of the radiation energy density and temperature.

B. **Stephen-Boltzmann Law**

The Stephen-Boltzmann law states that the total radiant thermal power emitted from a surface is proportional to the cubic degree of its absolute temperature. The law applies only to a blackbody, the theoretical surface that absorbs all incident thermal radiation. It provides a simple way to calculate the total power of blackbody radiation.

C. **Wien's law**

Wien's law states that at a certain temperature, the product of the absolute blackbody temperature and the peak wavelength λ corresponding to the radiation ability maximum is a constant, that is, $\lambda_m T = b$, where b is the Venn constant with a value of 0.002897 m · K. This law shows that the maximum of the absolute nobody increases. Venn's displacement law conforms not only to the short-wave part of the experimental curve of blackbody radiation but also to the entire energy spectrum of blackbody radiation, which is the maximum exploration of the blackbody radiation problem.

(2) **Gray body and the actual thermal radiation sources**

A. **Definition and characteristics of gray bodies**

Gray body is a concept in radiation heat transfer used to describe the process of thermal radiation, only a part of the thermal radiation reflects the rest of the object. The characteristic of the gray body is that its spectral absorption ratio is independent of the wavelength, that is, the absorption and reflection of the visible light band by the gray body are constant in each wavelength segment, and it does not have the characteristics of selective absorption and reflection. The radiation spectrum of the gray body is continuous, and the ratio of the radiation intensity of all the wavelength rays to the corresponding wavelength rays at the same temperature is equal to the constant. The shape of the radiation spectrum curve of the gray body is similar to that of the black body radiation spectrum curve, but its monochromatic radiation ability is not only smaller than the monochromatic radiation ability of the same wavelength, and the ratio of the two is not greater than the constant of 1. This characteristic allows the gray body to exhibit similar properties to the black body during thermal radiation but with weak radiation intensity.

B. **The difference between the actual thermal radiation source and the black body**

There are significant differences in the thermal radiation characteristics between the actual thermal radiation source and the black bodies. The black body is an idealized object that completely absorbs any external radiation at any wavelength without any reflection, that is, an object with an absorption ratio of one. The radiation spectrum of the blackbody is only related to its temperature, not to the material, and follows Planck's law of radiation. However, due to the influence of material, surface state, temperature, and other factors, the actual thermal radiation source and the black body are very different. First, the spectral properties of the actual thermal radiation source may be less ideal than the blackbody, where its radiation energy may be unevenly distributed at different wavelengths. Secondly, the absorption ratio of the actual thermal radiation sources may be less than 1, that is, they cannot fully absorb all the incoming radiation energy, but will reflect or transmit part of the energy. In addition, the radiation intensity of the actual thermal radiation source may also be affected by the temperature, surface state, and other factors. Therefore, in practical applications, we usually treat the actual thermal radiation sources as gray bodies or approximate gray bodies to better understand and predict their radiation properties. At the same time, we also need to choose the appropriate thermal radiation sources and radiation characteristics analysis methods according to the specific application scenarios and requirements.

(3) **Spectral characteristics of the thermal radiation sources**

The spectral properties of thermal radiation sources are used to describe the distribution of their radiation energy at different wavelengths. Here are two key points regarding the spectral characteristics of the thermal radiation source:

A. **The relationship between the spectral distribution and the temperature**

The spectral distribution of the thermal radiation is closely related to the temperature. According to Planck's law of radiation, the radiation intensity distribution is related to its temperature, and as the temperature increases, the radiation intensity increases over the entire spectrum. At room temperature, most of the objects mainly radiate infrared light. As the temperature increases, the spectral composition of the radiation changes. When the object is heated to about 500 °C, in addition to infrared, red light will be emitted; when the temperature rises further to 1500 °C, the object will emit all kinds of visible light, showing white; when heating up, the green and blue light will increase, and the object will appear green and blue. The energy radiated by the same object at a certain temperature is not distributed in different spectral regions. The higher the temperature, the higher the frequency in the spectrum corresponding to the most energetic radiation in the spectrum.

B. **The relationship between the peak wavelength and the temperature**

The peak wavelength of the thermal radiation spectrum also has a direct relation to the temperature of the object. According to Venn's shift law, the peak wavelength of the thermal radiation spectrum is inversely proportional to the temperature of the object, that is, the higher the temperature, the shorter the peak wavelength. This is also why high-temperature objects emit bluer light, because blue light has shorter wavelengths, while low-temperature objects emit more reddish light because red light is longer. Understanding the spectral properties of thermal radiation sources is very important for many applications, such as solar cells, thermal imaging, radiation temperature measurement, etc. All of these applications require accurate measurement and control of the spectral distribution and peak wavelengths of thermal radiation sources to achieve efficient and accurate energy conversion or temperature measurements.

(4) **Stability and reliability**

The stability and reliability of the basic characteristics of the thermal radiation source are the two key aspects of its practical application. Here is a detailed explanation of these two features:

A. **The importance of the stability of the thermal radiation sources**

The stability of a thermal radiation source refers to its ability to keep its radiation energy output constant over some time. This stability is crucial for many applications. For example, in the fields of measurement, sensing, and communication, the stability of the thermal radiation source directly affects the accuracy and reliability of the system. If the stability of the source of thermal radiation is insufficient, it may lead to measurement error, signal distortion, or communication interruption, which will affect the performance of the whole system.

B. **Factors affecting the stability of thermal radiation sources**

The radiation energy of a thermal radiation source is closely related to its temperature, so the temperature change is one of the main factors affecting the stability of a thermal radiation source. If the temperature of the thermal radiation source fluctuates during the operation, its radiation energy output will change accordingly, thus affecting the stability. The material properties of the thermal radiation sources can also affect their stability. For example, some materials are prone to changes such as oxidation, corrosion, or deformation at high temperatures, which may lead to changes in the radiation properties of thermal radiation sources, thus affecting their stability. The manufacturing process of the thermal radiation source can also affect its stability. If the manufacturing process is defective or insufficient, it may lead to structural instability or inconsistent performance of the thermal radiation source, thus affecting its stability. Environmental factors such as humidity, air pressure, and vibration may also affect the stability of the thermal radiation source. These environmental factors may lead to changes or interference in the performance of thermal radiation sources, thus affecting their stability.

To improve the stability of the thermal radiation source, a series of measures can be taken, such as optimizing material selection, improving the manufacturing process, and strengthening temperature control. At the same time, in the use of heat radiation sources, it is also necessary to pay attention to the impact of environmental factors on its stability and take corresponding measures for protection and compensation.

3. **The application of thermal radiation source in photoelectric detection**

(1) **Infrared detection system**

A. **Selection and application of infrared radiation sources**

Infrared radiation sources play a crucial role in photoelectric detection systems. Its selection and application should be determined according to the specific requirements of the detection system. Generally speaking, infrared radiation sources can be divided into natural infrared radiation sources and artificial infrared radiation sources. Natural infrared radiation sources mainly include various substances on the surface of the sun, the earth's surface, and its atmosphere. Among them, the sun is the most important natural infrared radiation source, and its radiation energy is widely used in various natural and artificial processes on the Earth's surface. Artificial infrared radiation source mainly refers to a variety of infrared light source devices, such as infrared laser, infrared LED, infrared thermal light source, etc. These devices can generate infrared radiation at specific wavelengths to meet the needs of different photoelectric detection systems. Infrared radiation sources have a wide range of applications, including night vision monitoring, remote sensing detection, medical diagnosis, communication transmission, and other fields. When selecting infrared radiation sources, their wavelength, radiation intensity, stability, and other factors need to be fully considered to ensure the performance and reliability of the photoelectric detection system.

B. **Basics of the infrared detection system**

Infrared detection system is a technology of target detection and information acquisition, based on the interaction principle of infrared radiation and matter. Its basic principle mainly includes the process of emission, propagation, absorption, reflection, and scattering of infrared radiation. The workflow of the infrared detection system mainly includes the following steps: (1) the infrared radiation source produces the infrared radiation of a specific wavelength. (2) Infrared radiation travels through the air or other media to reach the target object. (3) The target object absorbs, reflects, or scatters the infrared radiation, thus changing the energy and direction of the infrared radiation. (4) The infrared radiation after passing through the target object enters the infrared detector, and the detector converts the infrared radiation into an electrical signal. (5) After amplification and filtering, the electrical signals are input to the data processing system for image reconstruction, target recognition, and other operations. (6) Finally, the infrared detection system outputs the detection results, such as the temperature distribution map, the target position, the speed, and other information. The infrared detection system has wide application prospects in military, aerospace, environmental monitoring, biomedical, and other fields. A deep understanding of the basic principle and workflow of the infrared detection system can provide strong support for the development and application of infrared detection technology.

(2) **Temperature measurement and monitoring**

A. **Application of thermal radiation sources in temperature measurement**

Thermal radiation sources play an important role in temperature measurement. Thermal radiation is the electromagnetic radiation emitted by an object because of its temperature. The characteristics of this radiation are closely related to the temperature of an object. We can accurately understand the temperature state of the object by measuring the thermal radiation. In practical applications, heat radiation sources are mainly used to measure the temperature in a high-temperature environment. For example, in terms of furnace temperature measurement and temperature distribution measurement in the high-temperature furnace, the thermal radiation source can provide high-precision temperature data. In addition, in the space field, thermal radiation sources are also used to measure the temperature of the rocket engine to ensure that it works properly.

B. **Principles and technology of non-contact temperature measurement**

Non-contact temperature measurement refers to the measurement of the temperature of an object through some physical principle without touching the object under test. This measurement method has the advantage of not interfering with the object itself and can be measured in a dangerous environment. Among them, thermal radiation temperature measurement is a common non-contact temperature measurement technique. It determines the temperature of the object based on the relationship between the thermal radiation energy emitted by the object and the temperature of the object by measuring the thermal radiation energy emitted by the object. This method can realize long-distance, fast, and accurate temperature measurement, and is suitable for temperature monitoring in various high-temperature, corrosive, and dangerous environments. The key to thermal radiation temperature measurement lies in the selection and calibration of radiation temperature measuring instruments. Radiation temperature measuring instruments are usually composed of optical systems, photo detectors, signal processing circuits, and other parts. The optical system is used to collect the thermal radiation energy emitted by the object and focus it on the photo detector, which converts the collected radiation energy into an electrical signal and uses the signal processing circuit for amplification and filtering to obtain the temperature value of the object. To improve the accuracy of temperature measurement, the radiation temperature measurement instrument is calibration. During the calibration process, the radiation temperature measuring instrument is compared with the standard temperature source, and the parameters of the instrument are adjusted so that the output value matches the temperature value of the standard temperature source. In this way, the accuracy and stability of the radiation temperature measuring instruments can be ensured. In conclusion, the heat radiation source plays an important role in temperature measurement and monitoring. Through the non-contact temperature measurement technology, the temperature monitoring of high temperatures and dangerous environments can be realized to provide safety guarantees for various engineering applications.

(3) **Thermal imaging technology**

A. **Basic principles of the thermal imaging system**

The thermal imaging system [1] is a technique to convert the thermal radiation energy of an object's surface into a visual image. The basic principle is based on the theory of thermal radiation, that is, any object will radiate energy outward due to its temperature, and this energy radiation is closely related to the temperature of the surface of the object. The thermal imaging system detects this thermal radiation energy, converts it into electrical signals, and then uses image processing technology to form a temperature distribution image. The thermal imaging system usually consists of a thermal imaging sensor, signal processing unit, display, and other parts. Thermal imaging sensors are the core part of the system, which is to detect the thermal radiation energy on the surface of the object. The signal processing unit is responsible for magnifying and filtering the detected electrical signals to extract effective information. The display converts the processed signal into visual images for users to observe and analyze.

B. **The role of thermal radiation sources in thermal imaging technology**

In thermal imaging techniques, the thermal radiation sources are crucial. The characteristics of the thermal radiation source directly affect the detection effect of the thermal imaging system. The following are the key role of thermal radiation sources in thermal imaging technology: (1) providing stable thermal radiation energy: the thermal imaging system relies on the thermal radiation energy on the object's surface to work. Thermal radiation sources need to provide enough stable thermal radiation energy to ensure that the thermal imaging system can accurately detect and convert into electrical signals. (2) Difference in temperature distribution: the greater the difference in the temperature distribution of thermal radiation sources, the clearer the details that can be distinguished by the thermal imaging system. Therefore, thermal radiation sources play a role in providing differences in temperature distribution in thermal imaging techniques. (3) Band characteristics of the thermal radiation source: the band characteristics of the thermal radiation source determine the detection band of the thermal imaging system. Different thermal radiation sources have different band characteristics; therefore, thermal imaging systems need to be designed and adjusted according to the band characteristics of thermal radiation sources to achieve optimal detection results. Thermal radiation sources play a key role in providing stable thermal radiation energy, temperature distribution difference, and band characteristics in thermal imaging technology. These effects provide accurate and clear images of temperature distribution for thermal imaging systems, making thermal imaging technology widely used in many fields.

4. **Development trends of thermal radiation source technology**

(1) **High-temperature stability**

The high-temperature stability of heat radiation source technology is one of the key factors in its development. With the progress of science and technology, the stability of heat radiation sources in high-temperature environments has been significantly improved. This is mainly attributed to the application of new materials and the optimization of the structural design of the radiation source. In terms of high-temperature stability, current studies have focused on improving the oxidation resistance, corrosion resistance, and thermal stability of radiation sources. The improvement of oxidation resistance and corrosion resistance is mainly achieved by coating a high-temperature coating on the surface of the radiation source. This coating can effectively isolate the impact of a high-temperature environment on the radiation source, thus extending its service life. Moreover, by optimizing the structural design of radiation sources, they can be made more stable in high-temperature environments. For example, a cooling system is used to reduce the operating temperature of a radiation source and thus improve its high-temperature stability. The improvement of thermal stability mainly depends on the research and development of new thermal radiation source materials. These new materials have higher thermal conductivity, thermal expansion coefficient, and high-temperature resistance so that the radiation source can maintain a stable working performance in a high-temperature environment. At the same time, these new materials can also improve the radiation efficiency of radiation sources, to improve their high-temperature stability.

(2) **Wide spectral range**

With the development of thermal radiation source technology, thermal radiation sources with a wide spectral range have been widely used. Thermal radiation sources with a wide spectral range can emit thermal radiation over a wider wavelength range, thus meeting the needs of different application scenarios. The thermal radiation source in a wide spectral range is realized in two ways: by adjusting the composition and structure of the material, and by adopting the layered structure or microstructure by adjusting the structure parameters. Thermal radiation sources with a wide spectral range have wide applications in many fields, such as thermal imaging, laser technology, and infrared communication. The performance and efficiency of these applications can be improved.

(3) **Intelligence and integration**

Intelligence and integration are another important trend in the development of heat radiation source technology. By introducing intelligent and integrated technology, the performance and reliability of thermal radiation sources can be improved while reducing the cost. Intelligence is mainly reflected in the temperature control and regulation of heat radiation sources. By introducing advanced control algorithms and sensors, precise control of the temperature of the heat radiation source can improve its performance and reliability. Integration is mainly reflected in the combination of a heat radiation source and other functional modules to form a complete thermal management system. For example, thermal radiation sources are integrated with radiators, cooling systems, etc., to form an efficient thermal management system. The development of intelligent and integrated technology makes the heat radiation source more flexible and efficient in practical applications. For example, in electronic devices, the real-time monitoring and adjustment of the device temperature can be realized through an integrated thermal management system, thus improving the performance and reliability of the device.

(4) **Research and development of novel thermal radiation source materials**

The research and development of novel thermal radiation source materials is one of the important directions of heat radiation source technology development. Novel materials have unique physical and chemical properties to improve the performance and reliability of thermal radiation sources. At present, the research of new heat radiation source materials mainly focuses on the following aspects: first, the new high-temperature material, with high thermal conductivity, thermal expansion coefficient, and high-temperature resistance; second, the second is the new functional material, with special radiation properties, can improve the radiation efficiency of the heat radiation source; third, the new composite material, through the combination of different materials, realize the excellent radiation performance. The research and development of new heat radiation source materials can not only improve the performance of heat radiation sources but also broaden its application field. For example, in the space field, new thermal radiation source materials can improve the temperature control ability of spacecraft, thereby improving their performance and reliability.

## 2.3 Gas Discharge Light Source

1. **Overview of gas discharge light source**

(1) **Gas discharge light source**

A gas discharge light source [2, 3] is a light source made using the principle of gas discharge luminescence. During the gas discharge, the electrons accelerate in the electric field and then collide with the gas atoms, causing atomic excitation and ionization, which in turn generates light radiation. This kind of light source is widely used in lighting, display, spectral analysis, medical treatment, and scientific research.

(2) **Classification of the gas discharge light source**

The gas discharge light source can be divided according to different classification criteria. Here are some of the common categories:

A. **Classification by discharge voltage**

According to the different discharge voltage, the gas discharge light source can be divided into three categories: low-pressure discharge light source, high-pressure discharge light source, and ultra-high pressure discharge light source.

The low-pressure discharge light source is carried out at low pressure when the collision frequency between gas molecules is reduced and the ionization efficiency is relatively high. Common low-pressure discharge light source has fluorescent lamps and low-pressure sodium lamps. The fluorescent lamp is widely used in indoor lighting for its high efficiency, long life, and good color performance; the low-pressure sodium lamp is widely used in road lighting for its high light efficiency and long life.

The high-pressure discharge light source is carried out at higher pressure when the collision frequency between gas molecules increases and the ionization efficiency is relatively low. Common high-pressure discharge light source has high-pressured mercury lamps and high-pressure sodium lamps. High-pressure mercury lamp is widely used in industrial lighting due to its high light efficiency and low cost; high-pressure sodium lamp is widely used in road lighting due to its high light efficiency and long life.

The ultra-high pressure discharge light source is conducted at very high pressure when the collision frequency between gas molecules is very high and the ionization efficiency is relatively low. Common ultra-high pressure discharge light source has ultra-high-voltage xenon lamps and ultra-high voltage mercury lamps. Ultra HP xenon lamp is widely used in film shooting and stage lighting because of its high light efficiency and excellent light color performance; in large sports venues and outdoor lighting because of its high light efficiency and long life.

B. **Classification by discharge form**

According to the different forms of discharge, gas discharge light sources can be divided into two categories: arc discharge and glow discharge.

The working principle of the arc discharge light source is that the electrons in the gas are accelerated under the action of high voltage. When the electrons hit the gas atoms, they will ionize the gas atoms, thus producing arc light. These kinds of light sources include low-pressure gas lamps such as fluorescent lamps and low-pressure sodium lamps, as well as high-pressure gas discharge lamps such as high-pressure mercury lamps, high-pressure sodium lamps, and metal halide lamps. The fluorescent lamp is a use of low-pressure gas discharge principle, through the electron and the phosphor layer reaction, so that the phosphor fluorescence, thus producing light. The fluorescent lamp has the characteristics of high efficiency, low energy consumption, and long life. A low-pressure sodium lamp is the principle of low-pressure sodium vapor discharge luminous, the light is monochromatic, and the wavelength is about 589.0 nm, so it has high luminous efficiency and good color rendering. A high-pressure mercury lamp is the principle of using high-pressure mercury vapor discharge and luminescence. The light emitted contains ultraviolet light and visible light. Through the conversion of phosphor powder, you can get white light suitable for lighting. A high-pressure sodium lamp is the principle of high-pressure sodium vapor discharge light, the light is monochromatic, wavelength is about 589.0 nm, has high luminous efficiency, suitable for road lighting. Metal halide lamp uses the principle of high voltage metal halide steam discharge light emission. The light emitted contains ultraviolet light and visible light. Through the conversion of phosphor, you can get white light suitable for lighting.

The working principle of the glow discharge light source is that at a lower voltage, the electrons in the gas are excited, which will collide with the gas atoms, ionizing the gas atoms, thus producing a glow. Such light sources include neon lights and neon lights. Neon light uses the principle of low-voltage gas discharge, through the electronic reaction with the neon gas, making the neon gas glow, thus producing light. Neon lights are often used for outdoor advertising, decorative lighting, etc. Neon lamp is the use of low-pressure gas discharge principle, through the electron and neon gas reaction, make neon gas emit red light. A neon lamp is widely used in indicator lights, signal lights, decorative lighting, and other fields. All kinds of gas discharge light sources have their own unique characteristics and application fields, according to the actual needs of selection.

### C. **Classifying by light source type**

Mercury lamp is a kind of low-pressure mercury vapor discharge lamp, its light is mainly ultraviolet light, which is converted into visible light through phosphor. Mercury lamp is divided into straight tube type and compact type two kinds, widely used in lighting and ultraviolet curing and other fields. A sodium lamp is a high-pressure sodium vapor discharge lamp, the light is mainly yellow, suitable for road lighting and other need for large area lighting occasions. Sodium lamp has high luminous efficiency and long life, but the spectrum is narrow and the color reduction is poor. Xenon lamp is a kind of high-pressure xenon discharge lamp, the light emitted is close to the solar spectrum, the color reduction is strong, suitable for museums, operating rooms, and other lighting occasions with high color requirements. Xenon lamps have high luminous efficiency but a relatively short life. A hollow cathode lamp is a kind of low-pressure gas discharge lamp, and the light emitted is mainly infrared light and ultraviolet light, which is commonly used for spectral analysis and scientific research. The low cathode lamp has high luminous efficiency, but narrow spectral distribution. A deuterium lamp is a kind of high-pressure gas discharge lamp. The light emitted is mainly ultraviolet light, which is converted into visible light through phosphor powder. Deuterium lamps have high luminous efficiency and long life, but their narrow spectrum and color reduction is poor. All kinds of gas discharge light sources have their specific spectral distribution and application scenarios, and the selection of appropriate light sources needs to be comprehensively considered according to the actual needs.

(3) **Development history and current situation of gas discharge light source**

The development of gas discharge light sources has a long history. Since the end of the nineteenth century, with the progress of science and technology and the continuous improvement of people's demand for lighting, the gas discharge light source has been rapidly developed. Early gas discharge light source mainly includes carbon arc lamp, fluorescent lamps, etc. With the continuous progress of technology, new gas discharge light sources are constantly emerging, such as high-pressure mercury lamps, high-pressure sodium lamps, metal halide lamps, and so on.

At present, the gas discharge light source has been widely used in various fields. In the field of lighting, gas discharge light sources such as fluorescent lamps and high-pressure sodium lamps are widely used because of their advantages such as high efficiency, energy saving, and environmental protection. In the field of spectral analysis, gas discharge light sources such as mercury lamps and deuterium lamps are widely used because of their specific spectral distribution. In addition, the gas discharge light source also plays an important role in medical treatment, scientific research, and other fields.

The development of China's gas discharge lamp industry also shows a good trend. With the continuous progress of technology and the continuous expansion of the market, the product quality of China's gas discharge lamp industry is constantly improved, the cost is constantly reduced, and the market competitiveness has also been improved. It is expected that in the next few years, the market size of China's gas discharge lamp industry will continue to grow and make greater contribution to the development of the lighting industry.

2. **Basics of the gas discharge light source**

The basic principle of gas discharge light source involves the physical process of gas discharge, the luminous mechanism, and the working voltage and current characteristics. Here is a detailed explanation of these aspects:

(1) **Physical mechanism of gas discharge**

Gas discharge is a complex physical phenomenon that involves multiple physical processes such as ionization, excitation, collision, and thermal radiation. When a voltage is applied between the electrodes, the electric field strength gradually increases, and the gas atoms or molecules are subjected to the electric field force in the electric field, causing them to be stretched and ionized. In this process, electrons are released from atoms or molecules to form free electrons and positive ions. As the strength of the electric field increases, the electrons accelerate in the electric field and collide with other gas atoms or molecules, producing more electrons and ions, forming a gas discharge phenomenon.

(2) **Luminescence mechanism of the gas discharge light source**

The luminescence mechanism of a gas discharge light source is mainly to excite gas molecules or atoms through the collision of electrons in the gas to make them in the excited state. When these excited atoms or molecules return to the ground state, energy is released, producing light radiation. This luminescence mechanism is related to the gas type, pressure, current density, and other factors, and different gas discharge light sources produce different spectra. Specifically, the emission of gas discharge light source can be divided into two types: glow discharge and arc discharge. In the glow discharge phase, the electron and ion density in the gas is low, the current is small, and the light radiation generated is also weak, showing a faint glow. As the current increases, the gas discharge enters the arc discharge phase, when the electron and ion density increase and the current increases, the light radiation generated also increases, forming a bright beam.

(3) **Operating voltage and current characteristics of the gas discharge light source**

The operating voltage and current characteristics of a gas discharge light source are of important performance indexes. Generally speaking, the working voltage and current of the gas discharge light source are related to factors such as the structure of the lamp tube, the type and pressure of the filling gas, and the shape and material of the electrode. There is a relationship between voltage and current during gas discharge. When the voltage is low, the gas discharge is in the glow discharge phase, the current is small, and the lamp emits weak light. As the voltage increases, the gas discharge enters the arc discharge phase, the current increases, and the lamp emits bright light. However, if the voltage is too high or the current is too high, it may damage the lamp tube or shorten the service life. In addition, the starting voltage and maintenance voltage of the gas discharge light source are also important performance indicators. The starting voltage is the minimum voltage required to start the gas discharge, and the maintenance voltage is the voltage required to continuously stabilize the gas discharge. Different gas discharge light sources have different starting voltage and maintenance voltage characteristics.

3. **Basic characteristics of the gas discharge light source**

(1) **Spectral characteristics**

The spectral characteristics of a gas discharge light source refer to the wavelength distribution and intensity distribution of the radiated light. Different gas discharge light sources will produce different spectral characteristics, which mainly depend on the gas type, gas pressure, current density discharge mode, and other factors. For example, mercury arc light is a very effective UV source, most of its output is in the ultraviolet band (especially close to 254 nm); hydrogen and deuterium lamps can produce a strong continuous spectrum in the ultraviolet band, and the short-wave output is mainly limited by the light performance of the window.

(2) **Brightness and light intensity**

Brightness is the physical quantity describing the strength of the light flux emitted by a light source per unit area, while light intensity is the light flux emitted by the light source in a unit stereo angle in a certain direction. The brightness and light intensity of the gas discharge light source usually depend on factors such as the discharge current, gas type, and air pressure. For example, as the discharge current increases, the brightness and intensity of the light source usually increase.

(3) **Stability and the lifespan**

Stability refers to the degree to which the output light flux, spectral distribution, brightness, and other performance parameters of the light source remain stable during the long time of operation. The stability of the gas discharge light source is related to the lamp tube structure, filling gas type and air pressure, electrode shape, and material. Life refers to the time from the light source to the failure, which is affected by the tube material, manufacturing process, working environment, and other factors. The life span of gas discharge light sources can generally range from hundreds to thousands of hours.

(4) **Directionality and luminous angle**

Directionality refers to the distribution of the light emitted by the light source in space, and the angle of light emitting is the angular range describing the direction of the light source. The directionality and luminous Angle of the gas discharge light source depend on the shape of the lamp tube, the position of the electrode, and the discharge mode. For example, low-pressure gas discharge lamps such as fluorescent lamps usually have a large luminous angle, while a high-pressure gas discharge lamp may have a more concentrated luminous direction.

In general, gas discharge light sources have unique basic characteristics such as spectral characteristics, brightness and light intensity, stability and lifetime, direction, and luminous angle, which make them have a wide range of application values in different application scenarios.

4. **The application of gas discharge light source in photoelectric detection**

(1) **Gas discharge light source in spectral analysis**

Gas discharge light sources play important roles in spectral analysis. They can produce a spectrum with a specific wavelength distribution, used to analyze the elemental composition and structure of matter. For example, low-pressure gas discharge light sources such as low-pressure mercury lamps and low-pressure sodium lamps are often used as light sources for ultraviolet spectral analysis, while high-voltage gas discharge lamps can produce continuous spectra from visible to the range of ultraviolet light for broader spectral analysis.

(2) **Gas discharge light source in fluorescence detection and excitation**

In the field of fluorescence detection and excitation, the gas discharge light source is also indispensable. They can produce a high intensity of light radiation, used to excite the fluorescent material in the sample, and then analyze the properties of the sample by detecting the fluorescence signal. For example, fluorescent lamps, deuterium lamps, and other gas discharge light sources are often used in fluorescence microscopy, fluorescence spectrometers, and other equipment, used to observe the fluorescence characteristics of biological samples, and chemicals.

(3) **Gas discharge light source in imaging and lighting**

Gas discharge light source is also widely used in the imaging and lighting field. Because they can produce bright, uniform light radiation, they are often used in various lighting equipment, such as street lights, advertising light boxes, etc. In addition, in the field of imaging, the gas discharge light source is also often used as a light source, such as in fluorescence imaging, X-ray imaging, and other technologies, through the high-energy photons produced by the gas discharge light source to stimulate the sample or detector, and then obtain high-quality images.

(4) **Application of gas discharge light source in the calibration of photoelectric sensor and detector**

The gas discharge light source also plays an important role in the calibration process of the photoelectric sensor and the detector. Because the gas discharge light source can produce stable and controllable light radiation, it is often used as a calibration light source for testing the performance parameters of the photoelectric sensor and the detector, such as sensitivity, response time, etc. By comparing the output of the calibration light source with the actual measurement, it is possible to evaluate whether the performance of the photoelectric sensor and detector of the sensor meets the requirements, to ensure its accuracy and reliability in practical application.

Gas discharge light sources are widely used in the field of photoelectric detection, and they provide important technical support in the fields of spectral analysis, fluorescence detection and excitation, imaging and lighting, and photoelectric sensors and detector calibration.

5. **Development trend of gas discharge light source technology**

(1) **Improve the spectral purity and stability**

With the progress of science and technology, the gas discharge light source has increasingly high requirements in terms of spectral purity and stability. To achieve this goal, the researchers are working to optimize the gas discharge process and control the discharge parameters to obtain a purer and more stable spectral output. At the same time, the development of a new spectral filtering technology also provides the possibility to improve the spectral purity.

(2) **Develop new types of gas discharge light source materials**

To meet the requirements of different application scenarios for gas discharge light source performance, researchers are actively developing new gas discharge light source materials. These new materials may have the advantages of higher luminescence efficiency, longer life, and lower energy consumption, thus further improving the performance of the gas discharge light source. At the same time, the research of new materials also helps to solve the problems existing in the traditional gas discharge light source, such as low energy efficiency, large volume, short life, and so on.

(3) **Realize the intelligence and integration of the gas discharge light source**

With the continuous development of the Internet of Things, big data, and other technologies, the intelligence and integration of gas discharge light sources has become an important direction of future development. By combining the gas discharge light source with the intelligent control system, the remote monitoring, automatic adjustment, fault warning and other functions of the light source can be realized, thus improving the efficiency of use and safety. In addition, the integration of the gas discharge light source with other photoelectric detection technologies can form a more powerful photoelectric detection system to meet a wider range of application needs.

(4) **Explore the application of the gas discharge light source in the new photoelectric detection technology**

With the continuous emergence of new photoelectric detection technologies, the application of gas discharge light sources in these technologies also presents a new development trend. For example, in quantum communication, lidar, biological imaging, and other fields, gas discharge light sources may play an important role. By exploring the application of the gas discharge light source in the new photoelectric detection technology, its application scope can be further expanded to promote the development of related fields.

Gas discharge light source technology is developing towards improving spectral purity and stability, developing new gas discharge light source materials, realizing intelligence and integration, and exploring the application of new photoelectric detection technology. These development trends will help to promote the continuous progress of the gas discharge light source technology and the continuous expansion of the application fields.

## 2.4 Solid Luminous Light Source

### 1. Overview of the solid luminous light source

#### (1) Solid luminous light source

Solid light source, also known as solid light source (luminescence of solid), refers to the phenomenon when electromagnetic energy, electrical energy, mechanical energy chemical energy, etc., this energy is converted into light energy on the solid. The principle is based that after the solid material is excited by the external energy, the internal electrons transition to the excited state, and then release light energy in the process of returning to the ground state.

#### (2) Classification of solid luminescent light sources

There are many kinds of solid luminous light sources, and according to their luminous mechanism and application scenarios, they can be roughly divided into the following categories:

##### A. The phosphor luminescence

These light sources usually use certain solid materials to emit visible light under certain conditions (such as ultraviolet light). Phosphors are widely used in lighting, display, and other fields.

##### B. Light-emitting diode (LED)

LED [4] is a special semiconductor device, when the current passes through, its internal electrons compound with holes, releasing light energy. LED has the advantages of high efficiency, energy saving, and environmental protection, and has become the mainstream technology in the field of modern lighting and display.

##### C. Field-induced luminescent light source

This kind of light source uses an electric field to excite the electrons in a solid material, making it collide with the luminous center, and then produce light radiation. Field luminescent light source has the characteristics of low illumination and surface light source, which is suitable for the indication and lighting in special environments.

#### (3) The development trends of solid luminous light source

The development of solid light sources can go back to the end of the nineteenth century when scientists began to study the emission of solid matter. However, due to the limitation of technical conditions, solid luminous light sources have not been widely used for a long time. Until the early 1990s, with the rapid development of semiconductor technology, LED technology made a breakthrough, and solid light sources began gradually to become practical.

In recent years, with the continuous progress in materials science, nanotechnology, and other fields, the performance of the solid luminous light source has been significantly improved. For example, LED luminous efficiency has achieved a qualitative leap, from a few lumens/watts to hundreds of lumens/watts today, or even higher. In addition, the types of solid luminous light sources are also richer and richer, and the application field continues to expand, from the initial lighting, and display field expanded to biomedical, communication, military, and other fields.

At present, the solid luminous light source has become an indispensable part of the field of modern science and technology, and it is playing an increasingly important role in promoting scientific and technological progress and improving human life. In the future, with the continuous progress of technology and the continuous expansion of the application fields, the solid luminous light source will continue to usher in a broader development prospect.

2. **Basics of solid luminous light source**

(1) **Physical mechanism of solid luminescence**

The physical mechanism of solid luminescence mainly involves electronic transitions in solid matter. When the solid material is excited by external energy (such as electric energy, light energy, heat energy, etc.), its internal electrons will absorb this energy and transition from the ground state (low energy level) to the excited state (high energy level). When these electrons in the excited state return to the ground state, they release energy, which is emitted as photons, generating optical radiation. This process is called solid luminescence.

(2) **The working principle of the LED**

LED is a special solid light source, its working principle is based on the electronic structure and energy level structure of semiconductor materials. The LED is mainly composed of P-type semiconductors and N-type semiconductors, between which they form a PN junction. When a forward voltage is applied at both ends of the LED, the holes in the P region will move to the N region, and the electrons in the N region will move to the P region. Near the PN junction, the electrons recombine with the hole and release the energy, which is emitted as photons, resulting in optical radiation. The color and brightness of the LED depend on the type and structure of the semiconductor material it uses.

(3) **The luminous efficiency and quantum efficiency of a solid light-emitting light source**

The light emitting efficiency is one of the important parameters to describe the performance of a solid luminous light source, which represents the efficiency of the light source to convert the input energy into light energy. The luminescence efficiency can be expressed by the ratio of the optical power (or optical flux) to the input electric power. The higher the luminous efficiency, the higher the energy conversion efficiency of the light source, that is, the more energy saving.

Quantum efficiency is a description of the efficiency of a light source in the process of generating photons. It means how many photons the light source can produce for every photon or electron input. The higher the quantum efficiency, the higher the efficiency of the light source in the process of photon production, that is, the better the luminous performance of the light source. Quantum efficiency can be divided into internal quantum efficiency and external quantum efficiency. Internal quantum efficiency represents the ratio of the number of photons generated inside the light source to the number of photons input or electrons, and the ratio of the number of photons measured from outside the light source to the number of photons or electrons input. Because there are various losses (such as reflection, absorption, etc.) in the process of the light source, the external quantum efficiency is usually lower than the internal quantum efficiency.

3. **Application of solid luminous light source in photoelectric detection**

(1) **Application of solid luminescent light source in spectral analysis**

Solid luminescent light sources, such as LED, play a crucial role in the field of spectral analysis. Spectroscopic analysis is a technique to determine the composition and structure of matter by studying the optical properties such as absorption, emission, or scattering of light. Solid luminous light sources provide a stable and controllable light source, which is crucial for obtaining accurate spectral information. As a kind of solid luminous light source, LED has the advantages of high brightness, good luminous efficiency, and long life. In spectral analysis, LED can be used to excite the sample with light signals from specific components in the sample. By detecting these optical signals, we can obtain the spectral information of the sample, and thus infer the chemical composition, structure, and other information of the sample. For example, in chemical analysis, LED can be used to excite fluorescently labeled compounds, and the concentration of the compounds in the sample can be quantified by detecting the intensity and spectral properties of fluorescence. In biomedical research, LED light sources can be used to excite fluorescently labeled biomolecules or cells, which can provide insight into biological processes and disease mechanisms. Overall, the application of solid luminescent sources in spectral analysis greatly improves the accuracy and efficiency of analysis, providing a powerful tool for research in chemical, biological, environmental, and other fields.

(2) **Application of LED in fluorescence detection and excitation**

LED is also widely used in fluorescence detection and excitation. Due to the characteristics of high luminescence efficiency, good stability, and fast response speed, they are widely used in fluorescence microscopy, fluorescence spectrometers, and other devices to excite fluorescent markers and detect their fluorescence signals. In fluorescence microscopy, the LED acts as an excitation light source and can provide light at a specific wavelength to excite the fluorescent marker in the sample. Since the spectral properties of the LED can be precisely controlled, the wavelength of excited fluorescent markers can be precisely selected to obtain high-quality fluorescence images. The LED light source also has the advantages of long life and less heat, which helps to improve the operation efficiency and service life of the microscope. In the fluorescence spectrometer, the LED light source also plays an important role. The LED can provide a stable, tunable excitation light, and by measuring the fluorescence spectrum of the sample, it yields information about the composition and structure of the sample. The use of LED light sources also greatly reduces the operation cost and maintenance difficulty of the instrument. In biomedical research, the localization, tracking, and quantitative analysis of biomolecules, cells, etc. can be achieved by fluorescence labeling technology. The stability and controllability of the LED light source provide strong support for these studies. For example, in fluorescence quantitative PCR, the LED light source can be used to excite a fluorescent dye to accurately determine the concentration of the target DNA by measuring the change in the fluorescence signal.

(3) **The application of a solid luminous light source in imaging and lighting**

Solid luminous light sources are widely used in the field of imaging and lighting. The advantages of high brightness, long life, and fast response speed make the solid light source widely used in imaging and lighting. In the field of imaging, solid luminous light sources can be used to make high-brightness, high-resolution displays. For example, LED display and LCD both use the principle of a solid light source to achieve image display. In addition, solid luminous light sources can also be used to manufacture devices such as 3D scanners and barcode readers, which have important application value in logistics, manufacturing, and other fields. In the field of lighting, solid luminous light sources can be used to manufacture various lighting devices, such as LED lights, fluorescent lights, etc. This lighting equipment has the advantages of energy saving, environmental protection, and long life, and can be widely used in homes, businesses, industry, and other fields. In addition, the solid light source can also be used to make car lights, street lights, and other traffic lighting equipment to improve road safety.

(4) **The application of a solid luminous light source in the calibration of photoelectric sensors and detectors**

Solid luminous light sources play an important role in the calibration of photoelectric sensors and detectors. Solid light sources with stable light output and easily controlled light intensity can be used to calibrate a variety of photosensors with detectors. In the photoelectric sensor calibration, the solid luminous light source can serve as a standard light source to calibrate the sensitivity and dynamic range of the photoelectric sensor. By comparing the output signal of the photosensor with the light intensity of the solid luminous light source, the performance index of the sensor can be determined to ensure its accuracy and reliability. Furthermore, the solid light source can be used for the calibration of the detector. There may be performance differences in the fabrication of the detector, and calibration using a solid light source can ensure that the performance of the detector meets the predetermined requirements. During the calibration process, the solid light source can provide a stable optical signal, and the sensitivity and noise can be determined. In conclusion, solid luminous light sources play an important role in the calibration of photoelectric sensors and detectors. By calibrating by using a solid light source, it can ensure that the performance indexes of the photoelectric sensor and detector meet the predetermined requirements, improving their accuracy and reliability in practical applications.

(5) **Application of solid luminous light source in biomedical imaging and detection**

Solid luminescent light sources have many applications in the field of biomedical imaging and detection. In this field, solid luminous light sources such as LED are mainly used to excite fluorescent markers and detect their fluorescent signals, to image and detect specific molecules and cells in an organism. First, fluorescence imaging is a technique commonly used in biomedical imaging. By using a solid luminous light source such as LED as the excitation light source, the fluorescent markers in the organism can be effectively excited, and their fluorescence signal is detected. This imaging technique has the advantages of high sensitivity and high resolution and can be used to observe specific molecules in cells, receptors on the cell membrane, etc. Secondly, solid luminous light sources also play an important role in biomedical tests such as flow cytometry and immunoassays. Flow cytometry enables rapid analysis and sorting of cell populations by stimulating fluorescently labeled cells using solid luminescent sources such as LED and detecting their fluorescence signals. In immunoassays, a solid luminous light source can be used to detect the fluorescent signal of an antibody or antigen after binding to a fluorescent marker, thus enabling the detection of specific molecules. In conclusion, solid luminescent light sources have wide applications in biomedical imaging and detection. By using solid luminous light sources such as LED, we can effectively excite fluorescent markers detect their fluorescent signals, and realize the imaging and detection of specific molecules and cells in living organisms. This application not only improves the sensitivity and resolution of imaging and detection but also expands the scope and depth of biomedical research.

4. **The development trends of solid luminous light source technology**

The development of solid luminous light source technology, especially LED technology, is showing several significant trends. With the continuous progress of technology, the luminous efficiency and energy efficiency of solid light sources are constantly improving. For example, LED luminous efficiency has reached a new height and has the potential for further improvement. This helps to achieve more energy-efficient and more environmentally friendly lighting solutions. The application field of solid light sources is expanding. In addition to the traditional lighting and display fields, they are also widely used in biomedical science, natural resource exploration, chemical analysis, materials science, etc. In the future, with the further development of technology, the application field of solid luminous light sources will be further expanded. With the development of the Internet of Things, big data, and other technologies, the intelligence and integration of solid luminescent light sources have become an important trend. By integrating sensors, controllers, and other intelligent devices, solid light sources can achieve higher levels of functions, such as automatically adjusting brightness, color temperature, color, etc., to meet the needs of different scenarios. The development of new materials and structures will promote the further development of solid light source technology. For example, new materials and technologies such as nanotechnology and quantum dots (QDs) have been applied in the manufacture of solid luminescent light sources to improve their luminescent properties and stability. At the same time, the design of the new structure will also help to improve the energy efficiency and life of the solid light source. With the increasing attention on environmental protection in the world, the green environmental protection characteristics of solid light sources will get more attention. In the future, solid light sources will pay more attention to the use of environmentally friendly materials, the effective use of energy, and the treatment of waste.

The development trend of solid light source technology will revolve around higher luminous efficiency and energy efficiency, a wider range of applications, intelligence and integration, research and development of new materials and structures, and green environmental protection. These trends will promote the application of solid luminescent light source technology in lighting, display, biomedicine, natural resources exploration, and other fields to continue to broaden and deepen.

## 2.5 The Laser

1. **Laser overview**

(1) **Laser definition**

A laser [5, 6], called an "optical amplifier", is a device capable of producing and magnifying a laser. Laser, namely "Light Amplification by Stimulated Emission of Radiation", Chinese translated as "stimulated radiation light amplification". Through the principle of electron-stimulated radiation, the laser enlarges the light energy and produces highly consistent coherent light waves. This light wave has the characteristics of good monochrome, high coherence, and strong direction. The emergence of lasers has brought revolutionary changes to science and technology, industrial production, medical care, and other fields.

(2) **Classification of lasers**

According to the working principle and the type of laser-produced, lasers can be divided into the following categories:

A. **Gas laser**

A gas laser is a laser that uses a gas as the working matter. According to the working gas and excitation mode, the gas laser can be subdivided into many types, such as carbon dioxide laser, helium–neon laser, argon laser, etc. This kind of laser has the advantages of high power, good beam quality, and wide wavelength coverage, and has been widely used in industrial processing, medical treatment, scientific research, and other fields.

B. **Solid-state laser**

Solid-state laser is a laser that uses a solid medium (such as crystal, or glass) as the gain medium. Depending on the activated elements, solid lasers can be divided into doped and self-activated lasers. Doped lasers such as neodymium-doped laser, and self-activated lasers such as ruby laser. Solid-state laser has high output power and good stability and has important applications in military, medical, optical fiber communication, and other fields.

C. **Semiconductor lasers**

A semiconductor laser is a laser that uses a semiconductor material as an active medium. Semiconductor laser has the advantages of small size, lightweight, high electro-optical efficiency, and low working voltage, and is widely used in optical communication, computer interface, optical storage, and other fields.

D. **Fiber-optic laser**

A fiber laser is a laser that uses the optical fiber as a gain medium. According to the different gain mediums, it can be divided into single-mode fiber laser and multimode fiber laser. Fiber laser has the advantages of good beam quality, good heat dissipation performance, and compact structure, and has been widely used in industrial processing, medical treatment, scientific research, and other fields.

E. **Chemical laser**

A chemical laser is a laser that uses the energy released from a chemical reaction to generate a laser. According to the type of chemical reaction, it can be divided into photo gas chemical lasers and thermochemical lasers. Chemical laser has the advantages of high power and long life and has important applications in the military, rocket propulsion, and other fields.

(3) **Development trends of laser device**

Development of lasers began in the 1950s. In 1954, American scientist Theodore Maiman successfully developed the world's first ruby laser, marking the arrival of the laser era. Since then, the research and application of lasers have been rapidly developed worldwide. With the progress of material science, semiconductor technology, and optical fiber technology, there are more and more kinds of lasers, and their performance is constantly improving. At present, the laser is widely used in industrial processing, medical surgery, communication technology, scientific research, and other fields.

In China, the research and application of laser technology have also been greatly developed. It has not only made a series of breakthroughs in the field of scientific research but also achieved remarkable achievements in industrial application. China is trying to promote the innovation and industrialization of laser technology to meet the needs of national economic and social development. To meet the needs of industrial processing, medical surgery, and other fields, high-power, high brightness laser has become a research hotspot. With the development of semiconductor technology, the laser is smaller and smaller and more integrated. The combination of laser and numerical control systems, robots, and other technologies makes laser processing more intelligent and automatic.

2. **Basics of laser operation**

The physical basis of laser generation is mainly based on the theory of "stimulated radiation" proposed by Einstein in 1917. This theory states that an atom at a high level with a photon of its transition energy moves to low levels and releases a photon identical to the incident photon. This process is called stimulated radiation, which is the key to laser generation. In the laser, through the excitation of external energy, the atoms or molecules in the lasing medium (such as gas, solid, liquid, or semiconductor, etc.) absorb energy and transition to a high energy level, forming an excited state. When the atoms or molecules of these excited states return to low levels, photons are released and a laser light is generated by stimulated radiation.

3. **The application of laser in photoelectric detection**

(1) **Lidar**

Lidar is a radar system that uses a laser for detection and ranging. It determines information about the distance, speed, direction, and shape of the target by firing the laser beam to the target and measuring the time that the laser beam is reflected from the target. Lidar has the advantages of high precision, high resolution, and strong anti-interference ability, so it is widely used in autonomous driving, topographic mapping, environmental monitoring, and other fields [7–13].

(2) **Laser spectrum analysis**

Laser spectroscopy is a technique to analyze the composition and structure of matter using the spectral signal generated after its interaction with matter. Laser spectral analysis results in high sensitivity, high resolution, and nondestructive properties, so it is widely used in environmental monitoring, biomedicine, materials science, and other fields. For example, in environmental monitoring, laser spectroscopy can be used to detect concentrations of pollutants in the atmosphere; in biomedicine, it can be used for drug analysis and disease diagnosis.

(3) **Laser communication**

Laser communication [14–17] is a mode of communication using a laser beam for information transmission. It is transmitted by modulating the information to the laser beam and then using the high brightness and high directivity of the laser beam. Laser communication has the advantages of fast transmission speed, large bandwidth, and strong anti-interference ability, so it is widely used in high-speed data transmission, satellite communication, and other fields. For example, in optical fiber communication, a laser beam is used to transmit a digital signal; in satellite communication, the laser beam is used to achieve communication between a ground station and a satellite.

(4) **Laser imaging**

Laser imaging is a technique that uses laser beams for imaging. It is aged by shining a laser beam onto an object and then using the laser signal to reflect or scatter back from the object. Laser imaging has the advantages of high resolution, high contrast, and non-contact, so it is widely used in medical, industrial, and military fields. For example, in medicine, laser imaging can be used to diagnose and treat diseases; in industry, it can be used to detect material defects and measure object dimensions.

(5) **Laser microscopy technology**

Laser microscopy is a microscopy technique that uses a laser beam as a light source. It is aged by focusing the laser beam on tiny areas and then using laser signals reflected from objects or scattered back. Laser microscopy has the advantages of high resolution, high contrast, and non-destructive properties, so it is widely used in biomedicine, materials science, and other fields. For example, in biomedicine, laser microscopy can be used to observe the structure and function of cells and tissues; in materials science, it can be used to analyze the microstructure and properties of materials.

4. **Development trend of laser technology**

(1) **High power and high energy**

High power and high energy laser is an important direction for the development of laser technology. With the progress of science and technology and the demand for industrial manufacturing, the requirements for the power and energy of lasers are also increasing. High-power lasers can provide higher energy density, greatly enhancing the processing capacity in material processing, medical treatment, scientific research, and other fields. For example, under the application of high-power lasers, industrial processes such as cutting, welding, and punching can achieve higher efficiency and better processing quality. Similarly, high-energy lasers can provide enough energy to perform fine experimental operations in scientific research, such as particle physics and chemical reaction dynamics. In the medical field, high-energy lasers can be used for surgical cutting, tumor ablation, etc., and their precision and efficiency far exceed traditional surgical tools.

(2) **Ultrashort pulse and ultrafast modulation**

Ultrashort pulsed lasers and ultrafast modulation technology represent another important development trend of laser technology. Ultrashort pulse lasers, such as femtosecond and picosecond lasers, can achieve measurement and processing at ultra-high time resolution by generating a very short pulse time, bringing revolutionary progress to scientific research and industrial production. For example, in the biomedical field, femtosecond lasers can be used for precise cell cutting, while in the field of material processing, picosecond lasers can produce superfine patterns and structures. Ultrafast modulation allows the output of the laser to respond and modulate faster, which is crucial for applications that need to adjust laser parameters in real-time. In the field of communication, the ultrafast modulated laser can improve the speed and efficiency of data transmission; in scientific experiments, fast modulated laser can accurately control the experimental process to obtain more accurate experimental data.

(3) **Wide wavelength range and tunability**

Wide wavelength range and tunability are the two key characteristics in the development of laser technology, which are of great significance for scientific research and industrial applications. Lasers can provide wide wavelength range techniques, mainly by broadening the band structure of the emitting material or by adopting a composite emission mechanism. Such a wide wavelength range of lasers can emit light at a range of various wavelengths from ultraviolet to far infrared, providing powerful tools for material processing, biomedicine, communication technology, and other fields. A tunable laser enables continuous regulation of the wavelength, which typically depends on the structural tunability of the cavity inside the laser or the use of elements such as saturable absorbers. Fine regulation of the laser wavelength can be achieved by accurately controlling the parameters inside the laser, such as refractive index, dispersion, and gain medium. The combination of a wide wavelength range and tunability enables the laser to adapt to more application scenarios. For example, in the field of optical fiber communication, a wide wavelength range laser can support more channels, and improve the capacity and efficiency of data transmission; in the field of material processing, a wide wavelength range and tunable laser can meet the processing needs of different materials, realizing more refined cutting, welding and other processes.

(4) **Intelligence and integration**

The intelligent and integrated development of laser technology is an inevitable trend to adapt to the rapid development of modern science and technology. By introducing advanced control algorithms and artificial intelligence technology, intelligent lasers can realize real-time monitoring and automatic adjustment of laser working state. For example, through temperature control, pump power adjustment, and other means, the intelligent laser can automatically adjust the operating state during the environment change or working process to maintain the best performance. The integrated laser integrates multiple functional modules, such as pump source, cooling system, monitoring equipment, etc., into a single piece of equipment, which not only reduces the volume of the equipment but also improves the stability and reliability of the system. Integrated laser is easy to control, can simplify the workflow of laboratory or production line, and improve work efficiency. The combination of intelligence and integration makes laser development more efficient, more flexible, and easier to use. For example, in the industrial production line, the integrated laser can realize one-click operation, greatly reducing the operation difficulty and time; in the biomedical field, the intelligent laser can automatically adjust the parameters according to the experimental requirements to improve the accuracy and efficiency of the experiment.

The development trend of laser technology is high power and high energy, ultra-short pulse and ultra-fast modulation, wide wavelength range and tunability, and intelligence and integration. These trends will promote the application of laser technology in scientific research, industry, medicine, and other fields to broaden and deepen.

## References

1. Kruegle H (2017) Ch-20 Video technology overview. In: Fennelly FE (ed) Effective physical security. Butterworth-Heinemann, pp 347–385
2. Vohra D (2016) Fundamental characteristics. Apache HBase Prim Sept:3–8. https://doi.org/10.1007/978-1-4842-2424-3_1
3. Pueo B (2016) High speed cameras for motion analysis in sports science. J Hum Sport Exerc 11(1):53–73. https://doi.org/10.14198/jhse.2016.111.05
4. Ross JS et al (2014) Electrically tunable excitonic light-emitting diodes based on monolayer $WSe_2$ p-n junctions. Nat Nanotechnol 9(4):268–272. https://doi.org/10.1038/nnano.2014.26
5. Stenhoff S, Mills J (2024) Basic principles of lasers. Anaesth Intensive Care Med 25(2):133–137. https://doi.org/10.1016/j.mpaic.2023.10.013
6. Wilson J, Hawkes JFB (1987) Lasers: principles and applications. Prentice Hall
7. Wallace AM, Halimi A, Buller GS (2020) Full waveform LiDAR for adverse weather conditions. IEEE Trans Veh Technol 69(7):7064–7077. https://doi.org/10.1109/TVT.2020.2989148
8. Royo S, Ballesta-Garcia M (2019) An overview of lidar imaging systems for autonomous vehicles. Appl Sci 9(19). https://doi.org/10.3390/app9194093
9. Zhou X, Sun J, Fan Z, Li S, Lu W (2022) Research on detection performance improvement of polarization GM-APD LiDAR with adaptive adjustment of aperture diameter and spatial correlation method. Opt Laser Technol 155(Jan):108400. https://doi.org/10.1016/j.optlastec.2022.108400
10. Mizuno T, Ikeda H, Makino K, Tamura Y, Suzuki Y (2023) InGaAs Geiger-mode three-dimensional image sensor for flash LIDAR. J Soc Inf Disp 31:158–168. https://doi.org/10.1002/jsid.1196SPECIAL
11. Li Z et al (2015) Photon-counting chirped amplitude modulation lidar with 1.5-GHz gated InGaAs/InP APD. IEEE Photonics Technol Lett 27(6):616–619. https://doi.org/10.1109/LPT.2014.2386354
12. Liang Y et al (2022) Low-timing-jitter GHz-gated InGaAs/InP single-photon avalanche photodiode for LIDAR. IEEE J Sel Top Quantum Electron 28(2). https://doi.org/10.1109/JSTQE.2021.3109584
13. Sun X (2024) Review of photodetectors for space lidars. Sensors 24(20). https://doi.org/10.3390/s24206620
14. Yokota N, Yasaka H (2021) Spin laser local oscillators for homodyne detection in coherent optical communications. Micromachines 12(5). https://doi.org/10.3390/mi12050573
15. MohammadNejad S, Aghaei F (2020) Noise characteristics improvement of submicron InP/InGaAs avalanche photodiode for laser detection system. Opt Commun 455:124561. https://doi.org/10.1016/j.optcom.2019.124561
16. Laxmi V, Tyagi D, Parveen A, Singh L, Ouyang Z (2022) Silicon/graphene–ITO multiple heterojunctions and 1D PhC waveguide-based photodetection for mid-NIR IPE with high responsivity. Opt Commun 530(July 2022), 129142. https://doi.org/10.1016/j.optcom.2022.129142
17. Mak KF, Ju L, Wang F, Heinz TF (2012) Optical spectroscopy of graphene: from the far infrared to the ultraviolet. Solid State Commun 152(15):1341–1349. https://doi.org/10.1016/j.ssc.2012.04.064

# Chapter 3
# Single-Photon Detector

**Abstract** Single-photon detectors (SPDs) are critical components in quantum communication, imaging, and high-resolution sensing. These detectors operate by converting photon interactions into measurable electrical signals via the photoelectric effect. Various types of SPDs offer unique advantages in sensitivity, speed, and efficiency, enabling applications such as quantum key distribution (QKD), single-photon imaging, and laser ranging. Recent advancements in SPDs have significantly improved quantum communication systems. Challenges like low-temperature operation and scalability limitations are being addressed via innovations in materials, structural designs, and cooling technologies. Emerging THz SPDs show promise in terahertz microscopy, quantum radar, and long-distance communication. SPDs are poised to revolutionize areas ranging from quantum communication to medical imaging.

**Keywords** Photomultiplier tube · Quantum dot · Quantum communication · Single-photon detector · Avalanche photodiode · Lidar · Radar · Dark count rate · Afterpulse

## 3.1 Principle of Single-Photon Detector and Development Trends

1. **Overview of the single-photon detector**

(1) **Physical mechanisms of single-photon detection (SPD)**

A single-photon detector operates with ultra-high sensitivity to detect and count individual photons, converting optical signals into electrical signals. Its detection principle is based on the photoelectric effect, where the interaction of light with the detector alters the electronic state of atoms or molecules. This change is measured to detect photons.



S. Du and W. Yin, *Photodetection and Image Sensing Techniques*, Advanced Topics in Science and Technology in China 2, https://doi.org/10.1007/978-981-96-5479-6_3

(2) **Types of single-photon detectors**

A. **PMT**

The PMT [1–5] is a highly sensitive single-photon detector that combines the photoelectric effect with electron multiplication technology. It consists of a photocathode, a PMT, and an anode. When a photon strikes the photocathode, it generates electrons, which are then amplified through multiple stages in the PMT. The resulting electrons reach the anode, generating an electric current. PMTs are known for their high gain, high detection efficiency, and fast response, making them widely used in fields such as particle physics, optics, biology, and more.

B. **Avalanche PD**

An avalanche PD (APD) is a semiconductor detector that utilizes the avalanche effect to amplify electrons generated by light. When photons strike the APD, they create electron–hole pairs. These pairs undergo an avalanche process under the influence of an electric field, producing more electron–hole pairs. The resulting electrons are collected, generating a detectable current. APDs are known for their high gain and detection efficiency, making them ideal for high-speed, highly sensitive SPD.

C. **Superconducting nanowire single photon detector**

A superconducting nanowire single photon detector (SNSPD) is a device that detects single photons using superconducting nanowires. It operates based on the phase transition of the nanowires between superconducting and normal states. When a photon strikes the nanowire, it causes the nanowire to switch from a superconducting state to a normal state, generating an electric current. With its high gain, high detection efficiency, and fast response, SNSPDs are ideal for applications in quantum communication, quantum computing, and other fields.

D. **Quantum dot single-photon detector (QDSPD)**

A QDSPD is a single-photon detector based on QDs. It leverages the optical and electronic properties of QDs to detect single photons. When a photon strikes a QD, it excites electrons in the dot, generating an electric current. QDSPDs are known for their high detection efficiency, high gain, and adjustable detection wavelength range, making them well-suited for applications in quantum communication, quantum computing, and other fields.

E. **Novel single-photon detectors**

Additionally, several new detectors are currently under research and development. For example, single-photon detectors based on diamond nitrogen-vacancy (NV) centers exploit the optical and electronic properties of the diamond NV center. Other emerging detectors are based on Si nanostructures, metal nanostructures, and other novel materials. These new single-photon detectors offer distinct characteristics and are expected to play a significant role in fields like quantum communication, quantum computing, and more, by providing enhanced performance and versatility in various applications.

2. **Applications and development trends of single-photon detectors**

(1) **Quantum communication**

Quantum communication utilizes quantum entanglement for information transfer, with photons commonly serving as quantum state carriers. These photons transmit information via electromagnetic waves, often in the form of photonic state transmission. Over time, single-photon detectors have evolved, from traditional PMTs to semiconductor-based Si and InGaAs tubes, and more recently, superconducting single-photon detectors (SNSPDs), each suited to different applications.

Quantum Key Distribution (QKD), proposed in 1984, allows secure key exchange between two users over a potentially insecure quantum channel. In 2015, Boris Korzh introduced a QKD system based on indium gallium arsenide (InGaAs) single-photon detectors (SPDs), achieving secure key distribution over 307 km of ultra-low loss fiber (51.9 dB loss). SNSPDs, known for their high detection efficiency and low dark count rate (DCR), are increasingly used in QKD experiments. In 2005, Hadfield et al. demonstrated QKD using SNSPDs over 42.5 km, but the system had a high bit error rate due to limited performance. By 2007, Take Sue et al. achieved a 12.1 bps secure key rate over 200 km using SNSPDs, surpassing the performance of single-photon avalanche diode (SPAD) based systems. Since then, SNSPDs have significantly improved QKD distance and secret key rate, with a notable achievement in 2021 when an SNSPD-based TFQKD system proposed by Pittalugam reached a communication distance of over 600 km, improving the key rate by two orders of magnitude.

In 2020, Chinese researchers from the Chinese Academy of Sciences and Tsinghua University achieved a breakthrough in quantum communication by using a high-count-rate, low-noise single-photon detector and independent laser phase interference technology. This enabled a 500 km QKD in a real-world fiber network. In 2022, Guo Guangcan's team at the University of Science and Technology of China further improved the system, achieving secure transmission over 833 km using an SNSPD with low dark counts, reduced time jitter, and enhanced phase locking stability.

(2) **Laser ranging**

Laser ranging techniques, including time-of-flight (TOF), interferometry, and trigonometry, have been widely used for distance measurement. In recent years, the development of single-photon detectors has led to the emergence of a new method: single-photon laser ranging. This technique allows for longer-range measurements by detecting single photons, significantly enhancing the precision of distance measurements.

In 2000, Pellegrini et al. used a Q laser (25 MHz repetition frequency, 850 nm wavelength, 10 ps pulse width, and 10 pJ pulse energy) and a SiAPD to detect scattering signals from non-cooperative target surfaces. This setup achieved a detection distance of 50 m with a depth resolution of about 3 mm for 10 returned photons. In 2007, Warburton et al. from Heriot-Watt University were the first to use an SNSPD (with a jitter of about 70 ps) to achieve a 1 cm ground-to-ground resolution and a 4 mm depth resolution at a distance of 330 m with a 1550 nm wavelength laser.

In 2009, McCarthy et al. incorporated a scanning time-of-flight system and used SiSPAD (with a 180 μm active area and 400 ps jitter) to detect centimeter-level depth resolution for low-feature targets at distances beyond 325 m under daylight conditions. In 2016, Degnan et al. developed a frequency pulse laser with a 60 kHz frequency and sub-nanosecond pulse width. They used this system on an airborne platform to measure target reflection echoes, achieving sub-meter level measurement accuracy.

Domestic advancements in laser ranging have also seen significant progress in both accuracy and distance. In 2017, Nanjing University achieved a long-range measurement of 180 km using an SNSPD lidar system (1064 nm wavelength, with a DCR of less than 100 cps) under sea fog conditions. In 2020, Zhang Haitao and his team from the Yunnan Astronomical Observatory of the Chinese Academy of Sciences applied array detection technology and single-photon detectors to accurately measure the orbit altitude of a target 1000 km away, with a radar cross-section area of 0.045 $m^2$. In 2021, East China Normal University introduced a reference position in the single-photon ranging system, effectively reducing system delay drift. This improvement led to a photon flight time measurement accuracy of 0.5 ps, with a single-ranging distance of 65 μm.

(3) **Single-photon imaging**

With the continued development of photon-counting laser ranging technology, the time resolution of photon-counting laser ranging systems has significantly improved. China, in particular, has made rapid progress in single-photon imaging, establishing itself as a leader alongside the United States and other countries. Notable advancements include the 2011 development by the Shanghai Institute of Technical Physics, Chinese Academy of Sciences, of a photon-counting 3D imaging lidar system based on passive Q-laser technology, with a 13 kHz repetition frequency, 532 nm emission wavelength, 600 ps pulse width, and 1 μJ single-pulse energy. In 2013, the Shanghai Institute of Microsystem and Information Technology, Chinese Academy of Sciences, achieved 115 m laser ranging imaging with a 4 mm depth resolution using a 26.8 ps SNSPD device with very low jitter. That same year, the Buller team from Heriot-Watt University introduced the 1560 nm band SNSPD device into the time-of-flight deep imaging system, achieving centimeter-level resolution for object imaging at 1 km in daylight. In 2020, the American Jet Propulsion Laboratory (JPL), in collaboration with MIT, used a niobium-based SNSPD, achieving a temporal resolution of about 3 ps at 1550 nm, with sub-millimeter-level imaging accuracy. The University of Science and Technology of China (USTC) also achieved a significant breakthrough by completing a 3D imaging experiment over 45 km using a superconducting nanowire single-photon detector.

In 2021, Xu Feihu and his team at USTC utilized an InGaAs/InP-SPAD (20% photon detection efficiency (PDE), 210 ps jitter, 2.8 kHz DCR, and 32 ps time resolution) for non-horizon imaging and real-time tracking of hidden objects over 1.43 km. They also developed a near-infrared single-photon detector with a 1.4 ps time resolution and a 5 Hz DCR using pulse pump frequency up-conversion detection technology, achieving high-precision non-horizon imaging of millimeter-sized letters outside the view. This laid the foundation for future practical applications. Additionally, Xu's team introduced a compact coaxial single-photon lidar system with new noise suppression technology, achieving 3D imaging over 201.5 km with only 0.44 signal photons per pixel, marking a significant step toward practical, low-power lidar systems for ultra-long-range applications. As an emerging laser ranging method, single-photon laser ranging technology has seen significant progress in both long-distance laser ranging and single-photon laser imaging, paving the way for future applications in various fields.

(4) **Quantum correlation imaging**

Photon time-of-flight range imaging is a form of non-correlated imaging. In contrast, correlation imaging is based on the temporal and spatial correlation of photons. Quantum correlation imaging, also known as ghost imaging or two-photon imaging, is a novel imaging technique that uses the correlation of photocurrents to obtain object images. This method involves two light paths: one (signal light) interacts with the object and is directed to a barrel detector, while the other (reference light) does not interact with the object and is directed to a detector with spatial resolution capabilities. Early correlation imaging experiments used entangled photon pairs as the light source, offering unique features such as non-localized imaging and breaking the Rayleigh diffraction limit.

The development of ghost imaging began in 1995 when Shi Yanhua's group at the University of Maryland conducted the first ghost imaging experiment using entangled photon pairs generated by spontaneous parametric down-conversion. In 2002, Bennink et al. expanded this work by using random fluctuations of classical light sources instead of entangled photons. Since then, most ghost imaging research has relied on pseudothermal light sources. In 2005, the Chinese Academy of Sciences' Wu Ling'an team achieved the first lensless quantum imaging using a thermal light source with a short coherence time, which is easy to obtain and measure. Efforts to improve the speed and quality of ghost imaging have continued, with significant advancements in the following years.

In 2009, KATZO and colleagues used compressed sensing technology to enhance imaging quality. In 2010, the Italian Lugiato team introduced differential ghost imaging (DGI), which allowed for the measurement of an object's transmission function in absolute units, improving SNR's and significantly enhancing the quality of quantum imaging. In 2015, the Shanghai Optical Institute proposed a structured image reconstruction method that improved 3D ghost imaging lidar by incorporating orthogonality constraints. In 2016, Dai Qionghai's team from Tsinghua University used a spatial optical modulator to double the modulation frequency, achieving dynamic target imaging at 42 Hz with 8080 resolution. In 2018, Sun Mingjie and colleagues developed a computational ghost imaging system that used high-speed LED modules, enabling serial imaging at a 1000 frames per second rate with a 3232-pixel resolution, significantly improving the speed compared to previous systems. In 2021, Teruaki Torii's team proposed a time-division ghost imaging method, dividing the correlation function calculation into subunits to reconstruct high-quality images at a low SNR.

### 3. Single-photon detector calibration

#### (1) Calibration of the detection efficiency based on the standard detectors

Detection efficiency calibration based on a standard detector is performed using direct and indirect methods. The direct method involves calibrating the single-photon intensity with a low-noise detector, while the indirect method involves calibrating the initial optical power and decay values using a standard photo detector. The advantage of the indirect method is that it has a simpler system composition and a wider coverage band. However, its challenge lies in accurately calibrating the light attenuation across a large dynamic range. This method is highly versatile, and research institutions worldwide have significantly improved measurement accuracy by shortening the traceability of the source link, enhancing light source power stability, and using linear PDs.

Notable foreign research institutions, such as the German Federal Institute of Physical Technology (PTB), have contributed to this field (Fig. 3.1). The measurement uncertainty for the detection efficiency of single-photon detectors in the visible light range at PTB is approximately 0.2% (k = 1). Domestically, institutions like CLP 41 have also conducted related research. Their measurement uncertainty for detection efficiency is about 0.3% (k = 1), with the schematic diagram of the measurement device shown in Fig. 3.2.

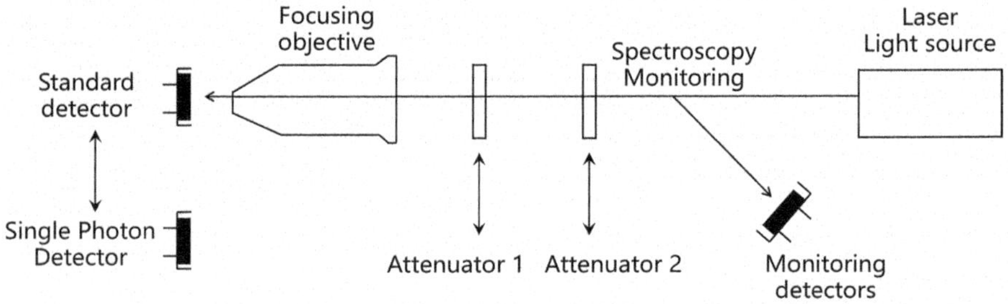


**Fig. 3.1** Schematic of the PTB single-photon detector detection efficiency calibrated using a standard detector

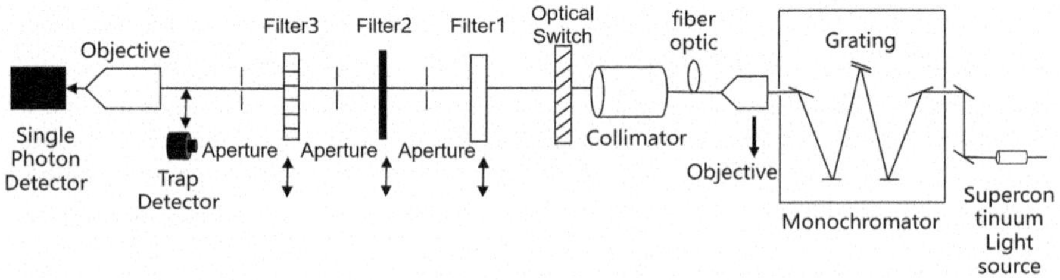


**Fig. 3.2** Schematic diagram of detection efficiency calibration device for SPD of CETC 41

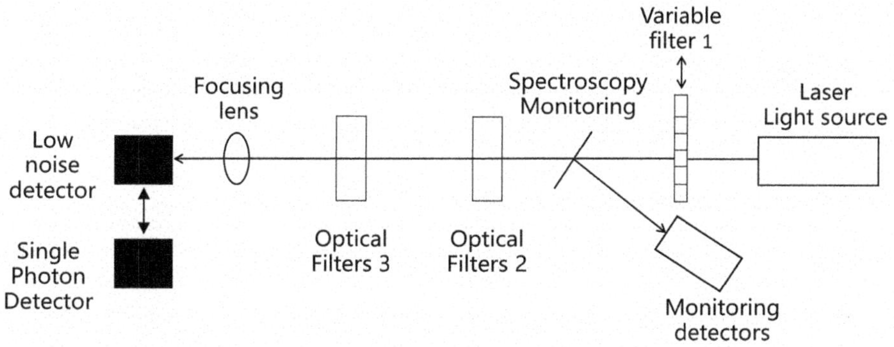


**Fig. 3.3** Schematic diagram of the experiment comparing German PTB with Czech CMI

With the improvement in the measurement accuracy of weak current signals, it has become possible to directly calibrate the single-photon intensity using a low-noise detector. Representative research institutions involved in this work include the National Institute of Standards and Technology (NIST) in the United States, the Czech Metrology Institute (CMI), and the National Metrology Institute of South Korea. In 2016, the German Federal Institute of Physical Technology (PTB) conducted a comparison experiment with CMI on the direct measurement of a standard single-photon detector. The results of this comparison, shown in Fig. 3.3, revealed that the $E_n$ value was less than 0.6, indicating good consistency between the two institutions' measurements.

(2) **The detection efficiency calibration based on the correlated photon pairs**

The calibration principle for detection efficiency based on correlated photon pairs involves generating pairs of correlated photons through the parametric down-conversion effect of a nonlinear crystal. The two photons are then transmitted along separate channels through spectral separation and detected by single-photon detectors. When photons are detected in one channel, the conservation of energy dictates that there must be an associated photon in the other channel. By measuring the counting signals from both single-photon detectors and incorporating the known counting signals and optical transmittance of the channel, the detection efficiency value can be calculated. This calibration method is traceable to fundamental quantum physics laws, distinguishing it from traditional traceability methods that rely on objective standards. As such, it is considered a "non-standard transfer" calibration scheme, aligning with the trend of metrological quantization and flattening.

International research institutions carrying out detection efficiency calibration based on correlated photon pairs include NIST (National Institute of Standards and Technology), the Italian National Metrology Institute, and NPL (National Physical Laboratory), with a detection efficiency uncertainty of approximately 0.2% ($k = 1$). Domestic institutions such as the Chinese Institute of Metrology, the Hefei Institute of Materials Science (Chinese Academy of Sciences), and CEC 41 have achieved detection efficiency uncertainties of less than 0.5% ($k = 1$). As a new calibration method, the correlated photon pair technique requires validation to ensure its measurement accuracy, and its effectiveness and feasibility are being demonstrated through studies by representative institutions like NIST and NPL. The $E_n$ value for the detection efficiency of standard detectors and correlated photon detectors in the visible light band is typically less than 1.

(3) **After-pulse probability calibration**

When a single-photon detector detects a photon, it triggers an avalanche process that generates a large number of carriers. While the majority of these carriers are released through the avalanche suppression circuit, a small fraction may be captured by low-level defects. After the avalanche is suppressed, these captured carriers are released and can trigger a secondary avalanche, leading to after-pulse counts, even in the absence of an incoming photon signal. This phenomenon results in miscounts, which can cause false detections. The after-pulse counts may cascade into multiple stages, known as multi-stage after-pulse counts. The ratio of the total after-pulse counts to the true photon counts is called the after-pulse probability ($P_{ap}$).

The key to measuring $P_{ap}$ lies in accurately detecting all the after-pulse counts resulting from one effective photon detection. In the case of SiAPD, $P_{ap}$ can be controlled within 1% under normal operation. These detectors typically work in free operation mode. To minimize $P_{ap}$, InGaAs/InP APDs are often operated in Geiger mode [6–13].

For measuring $P_{ap}$ in Geiger mode, Princeton Lightwave proposed a dual-timing measurement scheme in 2012. This method, as shown in Fig. 3.4, uses a laser with a frequency much lower than the detector's operating frequency, with a ratio of 1:128 between the laser frequency and the detector's frequency. The scheme records both the counting signal synchronized with the laser and the total counting signal output by the detector. This allows for comprehensive measurement of primary and higher-order after-pulse counts. The dual-timing scheme provides high measurement accuracy and is widely used in the calibration of $P_{ap}$.

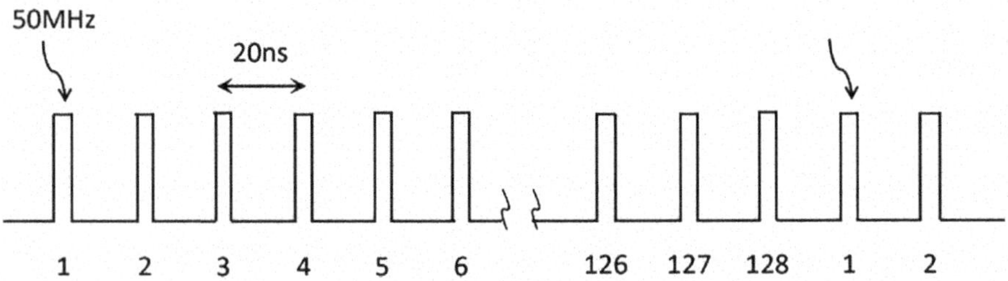


**Fig. 3.4** Schematic representation of the after-pulse probability measurements

## 3.2 Characteristics and Research Status of Single Photon Detector

1. **Novel single-photon detectors**

(1) **Superconducting Nanowire Single-Photon Detector (SNSPD)**

The SNSPD operates using ultrathin superconducting materials that detect single photons by creating localized "hotspots" upon photon absorption. These hotspots generate voltage pulses, enabling photoelectric conversion and SPD. SNSPDs boast nearly 100% theoretical system detection efficiency (SDE), surpassing the performance of traditional photo detectors and offering significant advancements in quantum technologies.

Notable developments in SNSPD include its integration with optical cavities and back mirrors in 2006, achieving SDEs of 57% at 1550 nm and 67% at 1064 nm. In 2017, the adoption of polycrystalline NbN nanowires with distributed Bragg reflection (DBR) mirrors enhanced absorption, resulting in an SDE of 90.2% at 1550 nm and 92.1% at 1.8 K. By 2019, a research team in China further optimized NbN-based SNSPDs by simultaneously improving detection response and optical absorption efficiency.

However, challenges persist, primarily the requirement for extremely low operating temperatures and large refrigeration systems, which hinder widespread application. Future research aims to address these limitations by integrating low-temperature technology, developing high-temperature superconducting materials, and creating SNSPD device arrays for broader use in advanced fields.

(2) **2D materials based APD**

Compared to traditional bulk materials, 2D layered materials exhibit exceptional properties such as self-passivating surfaces, strong light-matter interaction, adjustable Fermi levels, and mechanical flexibility. Additionally, their characteristics can vary depending on thickness, drawing significant attention from researchers. Among these, 2D graphene, a zero-band gap material, interacts with electromagnetic waves across a wide spectrum from ultraviolet to microwave bands, making it a versatile candidate for optical detection applications. However, its zero-band gap nature limits its suitability for optical detectors requiring a high SNR.

In contrast, 2D transition metal dichalcogenides (TMDs) like $WSe_2$ and $MoS_2$ [14–20], which possess specific band gaps, demonstrate excellent optical detection capabilities in the visible to near-infrared range. Another promising material is 2D black phosphorus (BP), with a direct band gap tunable from 0.3 to 2.0 eV in its monolayer form, making it ideal for infrared light detection. Advances in 2D materials technology have also inspired the development of novel single-photon detectors based on the APD mechanism [21], enabling high detection efficiency and single-photon counting via impact ionization.

2. **Research status of single-photon detectors**

(1) **Research on the free-running InGaAs/InP single-photon detector for real-time laser communication**

Recent advances in space exploration have spurred the development of long-distance communication systems like those involving satellites, the Moon, and Mars. Laser-based systems using single-photon detectors (SPDs) are crucial for high-sensitivity applications, including visible light communication (VLC), satellite-ground links, and deep-space optical communication. Notable efforts include NASA's 2013 Lunar Laser Communication Demonstration (LLCD) and subsequent projects like LCRD, DSOC, and DOCS. InGaAs/InPSPDs, operating in Geiger mode, are widely used due to their compactness, low cost, and reliability. SPDs are divided into SPAD arrays, which mitigate dead time and support high data rates (e.g., 14 Mbit/s at 1 photon/bit with an 88-size array), and single SPDs, which avoid array-related issues like crosstalk. In 2019, Wang et al. demonstrated a 2 Mbit/s transmission rate using a 20 MHz square wave, while in 2022, Tian et al. achieved sensitivities of – 57.2 dBm, – 53.42 dBm, and – 51.06 dBm at specific error rates using 1.8 GHz sinusoidal gating. To enhance SPD performance, optimized designs include dielectric-metal reflection layers and double Zn diffusion, leading to ~ 30% PDE, ~ 3 kHz DCR, and ~ 2.4% $P_{ap}$ at low temperatures. In real-time space laser communication, negative feedback avalanche diode (NFADs) achieved a transmission rate of 1 Mbit/s with a sensitivity of ~ – 69.6 dBm and a bit error rate of $1.1 \times 10^{-5}$.

A. **Device design and fabrication**

The InGaAs/InP SPD, shown in Fig. 3.5a, features separation absorption, gradient, charge, and multiplication heterostructures (SAGCM) and ladder PN junctions. It is fabricated using the metal–organic chemical vapor deposition (MOCVD) process, with an impurity concentration of $2 \times 10^{14}\,\text{cm}^{-3}$ in the epitaxial material, and the charge layer surface density is controlled to within 5%. The double Zn diffusion process is essential for creating the step PN junction, and the thickness of the multiplier area is controlled with ± 50 nm precision. The second Zn diffusion window defines an active region of approximately 25 μm in diameter. The doping concentrations in the gradient and charge layers adjust the electric field distribution in the absorption and multiplication layers. The electric field strength in the absorption layer is set between $1.1 \times 10^5$ and $1.4 \times 10^5$ V/cm to ensure the saturation drift rate of photo-borne carriers while avoiding avalanche breakdown. The doubling layer thickness is adjusted between 1 and 1.5 μm to achieve an electric field intensity greater than 400 kV/cm, which is required for high avalanche probability and reducing carrier transport time and timing jitter. The total depletion region thickness is approximately 3 μm. The p-type electrode is located on the top surface of the stepped zinc-doped zone, and the n-type electrode is located on the back side of the InP substrate, using Ti–Pt–Au and Au–Au–Ge–Ni alloys to minimize contact resistance.

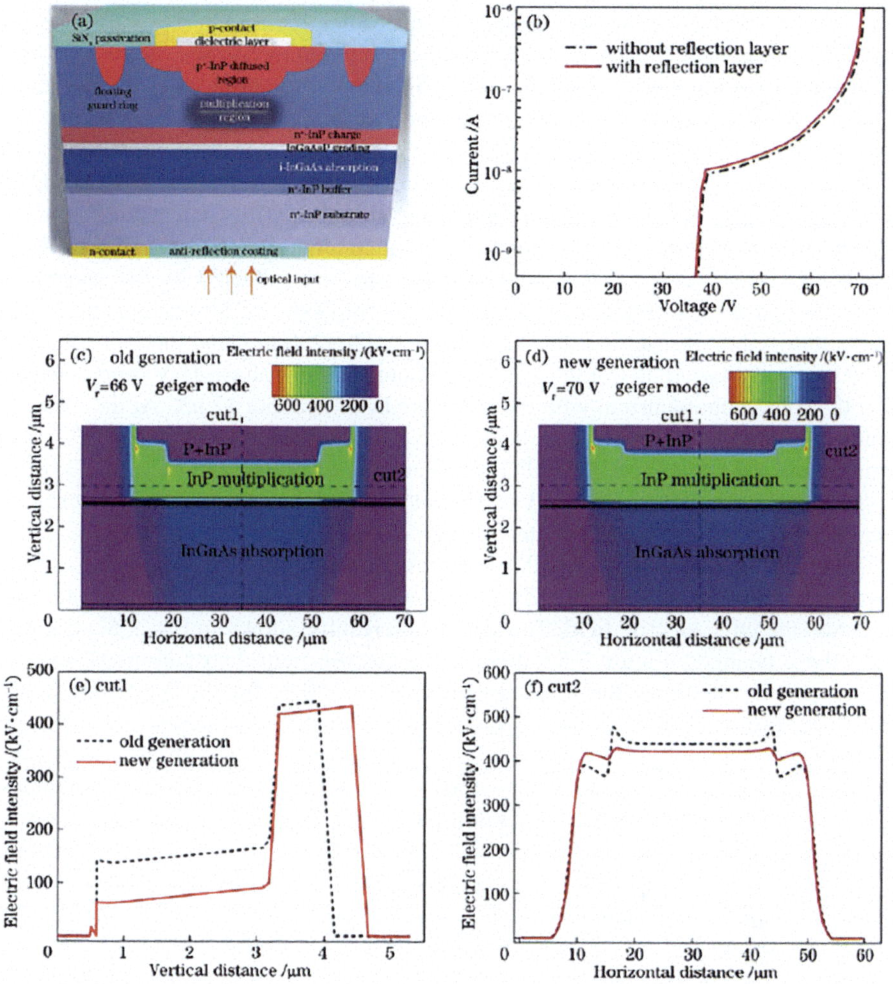


**Fig. 3.5** Structural design of InGaAs/InP SPD. **a** Schematic diagram of the typical In GaAs/InP SPD structure; **b** I-V curve of InGaAs/InP SPD with reflective layer and no reflective layer; **c** the electric field distribution of the existing InGaAs/InP SPD structure at 66 V; **d** the electric field distribution of the modified InGaAs/InP SPD structure at 70 V; **e** vertical electric field distributions against (c) and (d); **f** horizontal electric field distributions against (c) and (d)

Based on SAGCM and previous research, the semiconductor parameters have been further optimized. Key improvements include: (1) adding a high reflectivity medium-metal reflection layer to enhance the absorption of incident light and improve quantum efficiency, and (2) optimizing the double Zn diffusion process to adjust the depth and width of the step PN junction. This creates a uniform electric field in the SPD active area and inhibits edge breakdown, which reduces dark current, DCR, and $P_{ap}$. To support this optimization, the semiconductor process device simulation tool (TCAD) Silvaco, along with actual device test data, is used to select appropriate physical parameters, carrier mobility models, temperature correlation models, and collision ionization models for InP. This helps establish a comprehensive electrical simulation model, allowing accurate evaluation of the effects of design changes on parameters such as breakdown voltage, electric field, and avalanche probability.

To enhance the quantum efficiency of the device, a $SiO_2$ reflective layer is added to the top of the PN junction region in the chip structure. This increases reflectivity through the interference process between the semiconductor-dielectric and dielectric-metal interfaces. As a result, unabsorbed photons are reflected by the layer and reabsorbed in the absorption layer, improving the chip's overall absorption efficiency. The optimized dielectric film thickness is about 220–230 nm, and the reflectivity at 1550 nm is increased as in Fig. 3.5b. The photocurrent–voltage (I-V) curve shows that the source photocurrent with the reflective layer is about 12 nA. At the through voltage point, the SPD with the reflective layer gives 10.5 nA, compared to 9.4 nA without the reflective layer. The SPD quantum efficiency with the reflective layer is 87.5%, while without it, it is 78.3%. The addition of the reflective layer increases quantum efficiency by approximately 10 percentage points.

The improved InGaAs/InP SPD structure increases the charge in the charge layer and optimizes the double Zn diffusion process to reduce dark current and suppress the edge electric field. The simulation results, shown in Fig. 3.5c–f, highlight these changes. Figure 3.5c, d show the electric field distributions at reverse biases ($V_r$) of 66 V for the existing structure and 70 V for the improved structure. The improved design features a shallower zinc diffusion depth and a thicker doubling zone. Figure 3.5e compares the vertical electric field, with the new device showing a decrease in intensity from ~ 460 to ~ 420 kV/cm, reducing dark current by lowering defect-assisted tunneling. Additionally, the charge layer increase reduces the absorption layer's electric field from $\sim 1.5 \times 10^5$ to $\sim 1.0 \times 10^5$ V/cm, mitigating dark current from interband tunneling. Figure 3.5f shows the horizontal electric field distribution in the multiplication area, where the edge electric field is flatter, and effectively suppressing strong edge fields in the new structure.

Figure 3.6 presents the dark current and photocurrent test data for InGaAs/InPSPD at an optical power of 1 μW. As the temperature decreases from 300 to 213 K, the breakdown voltage ($V_{br}$) of the device drops from ~ 71 to ~ 62 V, with a temperature coefficient of approximately 105 mV/K. At a reverse bias of 1 V, the dark current at 300 K is ~ 0.3 nA and at 213 K is ~ 18 pA. The DC gain at 213 K, as shown in the DC gain curve in Fig. 3.6, is about 200, which exceeds the test value of the commercial InGaAs/InPSPD (PGA-300).

B. **Single-photon detection**

A single-photon calibration protocol was used to characterize the InGaAs/InP SPAD samples, with the SPAD operating in a 1.25 GHz sinusoidal gated mode. Time-correlated single-photon counting (TCSPC) technology was applied to test performance parameters such as PDE, \DCR, and $P_{ap}$. The unit SPAD module, shown in Fig. 3.7a, integrates the SPAD chip, circuit control, and temperature control functions, with the circuit principle depicted in Fig. 3.7b. This module supports optical

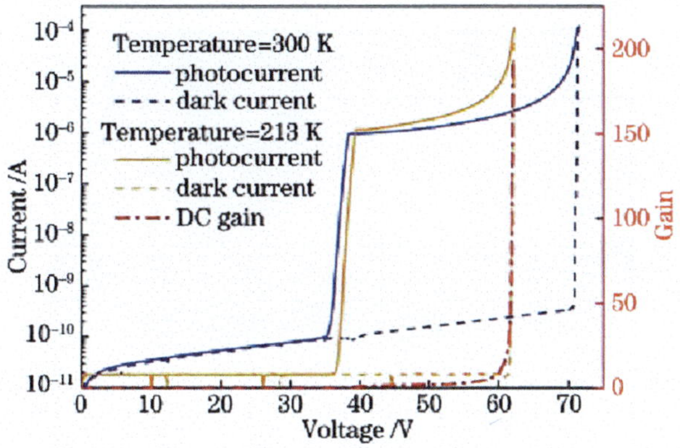


**Fig. 3.6** Curves of the tested photocurrent, dark current and calculated DC gain with reverse bias for an optical power of 1 μW at different temperatures

interfaces such as free space optical windows, FC/PC ports, multimode fibers, and is equipped with parameter test software developed using Labview (Fig. 3.7c). The test settings include a temperature range of 300–200 K, a 1.25 GHz sinusoidal gated quenching circuit with a maximum amplitude of 30 V, a 1550 nm pulse laser with a 200 ps full width at half maximum (FWHM), an attenuator set to achieve one average photon per pulse, and a dead time of approximately 88 ns to eliminate the influence of after-pulses.

Figure 3.8a shows the PDE of the device at different temperatures and overbias voltages ($V_X$). At 300 K, the PDE increases from 0 to 8.7 V, reaching a maximum of about 40%. At lower temperatures (213–243 K), the PDE also increases with $V_X$ in the 0–6.5 V range. When $V_X$ is constant, the PDE increases as the temperature decreases, due to the higher collision separation rate and avalanche probability. Figure 3.8b, c show the variations of DCR and $P_{ap}$ with PDE. Both DCR and Pap decrease significantly with temperature, while increasing rapidly with PDE. Overall, the device demonstrates excellent single-photon performance with ~ 30% PDE, ~ 3 kHz DCR, and ~ 2.4% Pap.

C. **Laser communication experiment**

To overcome the complexity of the SPAD circuit in gated mode, a simpler free-operation NFAD was developed by integrating a large resistance onto the SPAD chip. This device achieved 40% PDE, 2300 $s^{-1}$ DCR, 8% $P_{ap}$, and 49 ps timing jitter. The NFAD was then used as a receiver in a real-time laser communication system, with bit error rate (BER) testing to verify the system's feasibility and stability. The bit error rate test setup, shown in Fig. 3.9, provides a 1 Mbit/s service rate and 2 Mbit/s line rate for 4 pulse position modulation (PPM) encoded signals. The system consists of the sending terminal (Tx), which encodes the signal, and the receiving terminal (Rx), implemented in a field-programmable gate array (FPGA, XC7K325TFFG900). The test begins when the pseudo-random binary sequence (PRBS) from the PC drives a 1550 nm laser through the FPGA, producing an optical pulse signal with a pulse width range of 4–39 ns and a peak power of 75 mW. This optical signal is then attenuated and coupled to the NFAD through fiber. Thanks to the excellent performance of

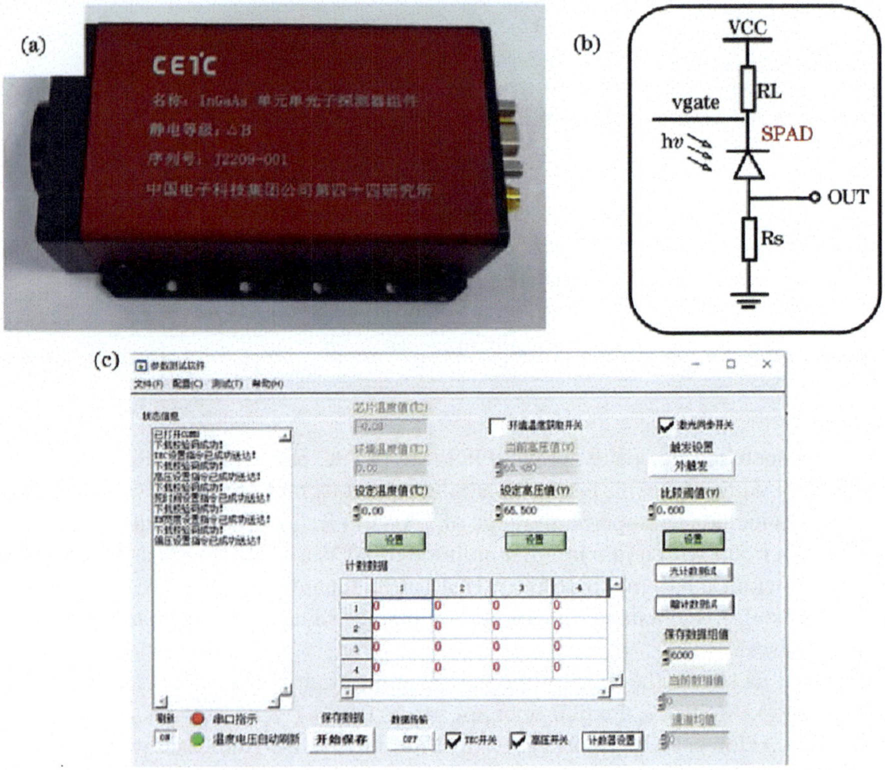


**Fig. 3.7** SPAD prototype and working principle. **a** SPAD module sample; **b** SPAD circuit diagram; **c** module control software interface

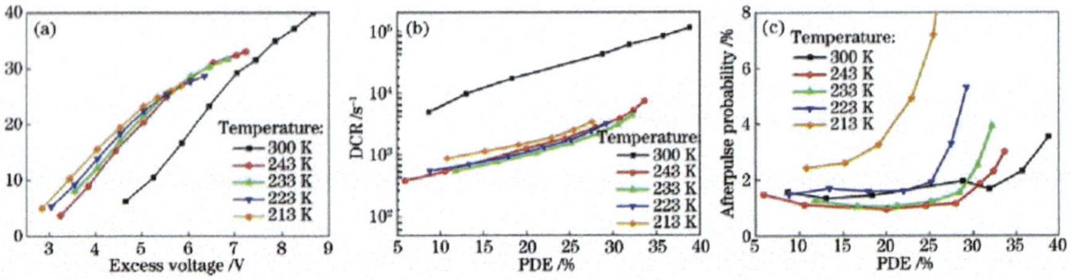


**Fig. 3.8** Plot of PDE with over bias tested at different temperatures, plot of DCR with PDE and plot of Pap with PDE tested at different temperatures

NFAD in free operation mode, real-time electrical signals are output for 2 Mbit/s transmission. The receiving terminal samples the output electrical signals from the NFAD, synchronizes, and decodes the 4 PPM signal, which is sent to the PC for real-time bit error rate (BER) calculation.

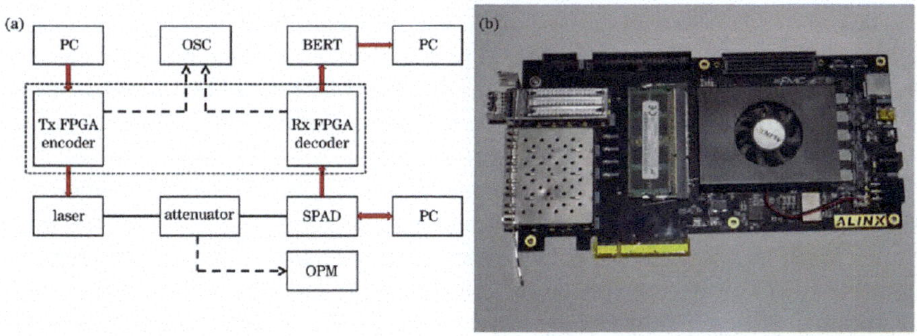


**Fig. 3.9** SPAD laser communication experiment. **a** Schematic diagram of laser communication experimental device; **b** FPGA module integrating sending and receiving functions

During the experiment, the NFAD was controlled through the upper computer, with the operating conditions set as follows: operating temperature of T = – 10 °C, operating voltage of V = 64.2 V, and a dead time of t = 150 ns. The laser was modulated using a 4 PPM scheme of "1000" at a transmission rate of 1 Mbit/s. The laser was turned on when sending a high level (1) and turned off when sending a low level (0), producing a laser output signal waveform. The signal was then attenuated, and the attenuated light was measured with an optical power meter, resulting in an incident light intensity of – 60 dBm. This optical signal was transferred from the optical power meter to the NFAD for testing. The test results were processed by an FPGA for decoding, and the decoding signal was sent to an oscilloscope for signal observation, as shown in Fig. 3.10a. After each modulation signal's high level (1), the system achieved signal synchronization, allowing the 4 PPM scheme to function correctly. Next, a random PRBS with a 1 Mbit/s bit rate was used for a bit error test. The NFAD's operating conditions remained unchanged, but the incident light power decayed to – 69.6 dBm, as shown in Fig. 3.10b. A certain signal loss was observed compared to the synchronous signal, leading to some error codes. Finally, the signal was connected to the upper computer for bit error rate (BER) calculation, yielding a BER of $1.1 \times 10^{-5}$.

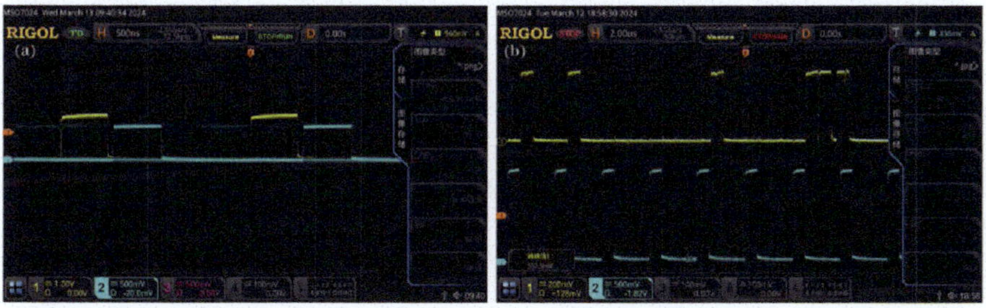


**Fig. 3.10** The NFAD BER test. **a** The test results obtained with 1000 manual codes and **b** the test results

The error code observed in the experiment can be attributed to two main factors related to the SPAD process, which is a probabilistic event. One potential cause is that the photon arrival moment did not trigger the avalanche effect, resulting in no light count being produced. The second cause is the occurrence of an avalanche event without the arrival of a photon, leading to an unintended light count. The first case corresponds to the single-photon detector's (SPD) insufficient detection efficiency, while the second case is linked to the detector's dark count and $P_{ap}$ being too high. As discussed in Sect. 3.3, these factors are closely related to the SPD's operating temperature (T), operating voltage (V), dead time (t), and input optical power ($P_{in}$). By optimizing these parameters, the system achieved a bit error rate (BER) of 1.1 $\times$ $10^{-5}$ under the conditions of T = − 10 °C, V = 64.2 V, t = 150 ns, and $P_{in}$ = − 69.6 dBm. This result verifies the feasibility of using the single-photon detector in free-operation mode for space laser communication, providing a solid foundation for further research on high-speed single-photon detector arrays.

The optimized InGaAs/InPSPD is thoroughly described in terms of its design, production, and testing. Key elements include: (1) the addition of a high-reflectivity medium-metal reflectance layer to the device structure, along with the optimization of the dual Zn diffusion process, which adjusts the ladder PN junction depth and width. This improves SPD absorption of incident light, enhances quantum efficiency, and creates a uniform electric field in the center of the SPD active region, helping to suppress edge breakdown. This, in turn, reduces dark current, DCR, and $P_{ap}$; (2) In the single-photon performance test, with high-frequency sinusoidal gating at 1.25 GHz, at 225 K and an over-bias of 6 V, the SPD achieved a detection efficiency of 30%, a DCR of 3 kHz, and a $P_{ap}$ of 2.4%. A single NFAD, prepared from the high-performance SPAD, was then applied in a real-time space laser communication system. The system achieved a bit error rate (BER) of 1.1 $\times$ $10^{-5}$ at a 1 Mbit/s transmission rate and demonstrated a sensitivity of approximately − 69.6 dBm. This demonstrates that the NFAD can significantly improve system performance for practical applications requiring near-infrared asynchronous SPD and real-time space laser communication.

(2) **High-performance superconducting phase transition single-photon detector**

Astronomical observation research has evolved into full-band astronomy, allowing for more dimensional insights into research objects through multi-band collaboration. Visible-light and near-infrared astronomy have seen significant progress, especially in areas like dark matter, dark energy, first-generation stars, galaxies, and supermassive black holes. Many large optical/infrared telescopes, including the Thirty Meter Telescope, are under construction, highlighting the scientific significance of these observations. Traditional optical telescopes use CCD cameras, which have nearly 100% quantum efficiency in the visible band but suffer from increasing readout noise at high speeds, making it difficult to observe fast, transient sources. As the wavelength extends into the infrared, HgCdTe detectors offer higher sensitivity, as seen with the James Webb Space Telescope (JWST) and its observations of Jupiter.

The superconducting phase transition edge detector (TES) has gained attention for its high detection sensitivity, photon number resolution in visible/near-infrared bands, and high energy resolution in X-rays. Compared to semiconductor detectors like Si-based CCDs, InSb, and HgCdTe, TES detectors offer higher detection efficiency, faster response, lower dark counts, and better energy resolution, making them ideal for time-domain astronomical observations. Stanford University has developed an imaging spectrometer based on TES, successfully observing the Crab pulsar's energy spectrum and time-varying characteristics. Later, a 32-element TES detector array was integrated into an adiabatic demagnetization refrigerator, supporting future astronomical observations.

In exoplanet research, searching for biological signals in a planet's atmosphere (e.g., $H_2O$, $O_2$) is crucial. This requires high energy resolution detectors across the ultraviolet–visible-near-infrared range (400–1800 nm). Superconducting TES detectors are expected to be pivotal, especially considering weak signals in exoplanet research. The Origin Space Telescope (OST), part of NASA's 2020 astrophysics decade plan, will cover the full infrared spectrum and achieve over 1000 times the sensitivity of previous telescopes. The U.S. is developing a 10,000-magnitude superconducting TES array for OST with extremely low noise-equivalent power.

The National Institute of Standards and Technology has developed a superconducting TES single-photon detector using 20 nm tungsten film, achieving 95% quantum detection efficiency at 1560 nm and an energy resolution of 0.29 eV. Japan's titanium superconducting TES detector at 850 nm achieves 98% detection efficiency and 0.42 eV energy resolution. In collaboration with Tsinghua University and the University of Science and Technology of China, a superconducting TES single-photon detector was developed based on Al/Ti bilayer membrane, with an effective response time of 3.9 μs. The Chinese Purple Mountain Observatory successfully developed a superconducting TES detector array for 350 GHz that meets ground observation sensitivity limits, extending to optical/near-infrared bands with an 80% system detection efficiency and the ability to distinguish photons at 1550 nm.

### A. **Superconducting TES single-photon detector**

The superconducting phase TES, as shown in Fig. 3.11a, operates as a heat detector consisting of an absorber (with heat capacity C), a weak thermal connection (thermal conductivity G), and a heat sink (at temperature Tb). When an incident photon with energy $E_\gamma$ is absorbed, the electron temperature (T) of the superconducting TES detector increases by $\delta T = E_\gamma / C$. The structure of the superconducting TES SPD in the visible/near-infrared band is shown in Fig. 3.11b.

### B. **Characteristic regulation of superconducting titanium films**

A superconducting titanium film was fabricated on a monocrystalline Si substrate by electron beam evaporation to measure its stress, surface resistance ($R_S$), and critical temperature ($T_C$). All samples exhibited compressive stress, which gradually decreased with increasing deposition rate. When the deposition rate exceeded 1 nm/s, the compressive stress remained constant. Similarly, the surface resistance decreased with increasing deposition rate, stabilizing after a deposition rate of 0.1 nm/

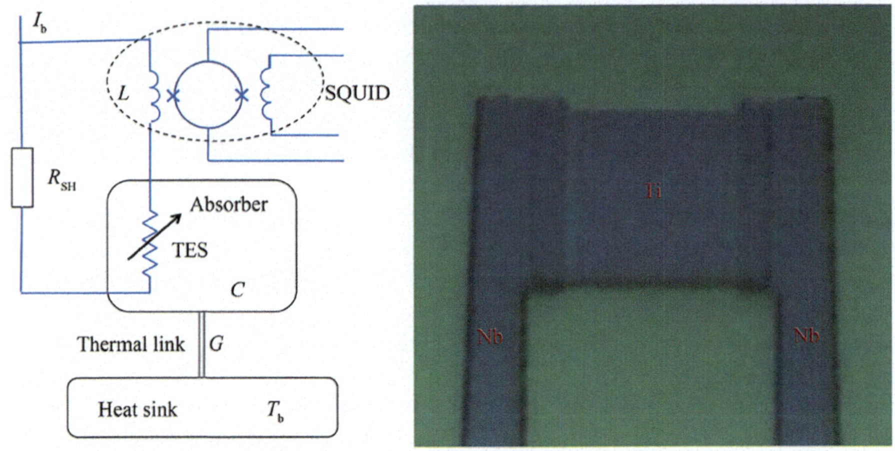


(a) Diagram of superconducting TES (b) Photo of optical superconducting TES

**Fig. 3.11** Schematic and photograph of the superconducting TES detector

s. Therefore, a deposition rate of 1–1.5 nm/s was chosen when developing the superconducting TES single-photon detector.

The $T_C$ of the superconducting titanium membranes showed a strong correlation with thickness (d) and surface resistance ($R_S$). As the thickness increased, the $T_C$ rose and $R_S$ decreased. However, other factors influenced these parameters, resulting in some data dispersion. A model proposed by IVRYY was used to describe the relationship between T and $R_S$: $dT_C = aR_S^{-b}$ where $a = 228.951$ and $b = 0.961$. The fitted curves matched the measured results well, showing that $T_C$ decreases with increasing $R_S$. The relationship allows for predicting the $T_C$ of titanium films as: $T_C = \frac{228.951}{R_S^{0.961}}$ which helps improve both $T_C$ and the consistency of the preparation process.

For the prepared superconducting TES single-photon detector, to accurately regulate $T_C$ and achieve high energy resolution, a post-baking process was proposed. The sample is placed in a conventional oven and baked at a set temperature. This high-temperature baking process oxidizes the surface of the titanium film, reducing its effective thickness and consequently lowering $T_C$. The effects of baking temperature ($T_{baking}$) and baking time ($t_{baking}$) on $T_C$ were further investigated. As shown in Fig. 3.12, $T_C$ of both Ti membranes and TiAu bilayers decreased with increasing baking temperature and time. For further quantification, with a fixed baking temperature of 120 °C, the change in $T_C$ was studied as a function of baking time ($t_{baking}$).

$T_c(t_{baking}) = T_{C0} - c \times \log(t_{baking})$ Where, c is the fitting parameter. Fixed the baking time ($t_{baking} = 10$ h), the change $T_C(T_{baking}) = T_{C1}\left[1 - \left(\frac{T_{baking}}{T_0}\right)^2\right]$.

When the titanium film is baked at high temperatures, oxidation occurs on the surface of the titanium film, which penetrates deeper into the material, resulting in a

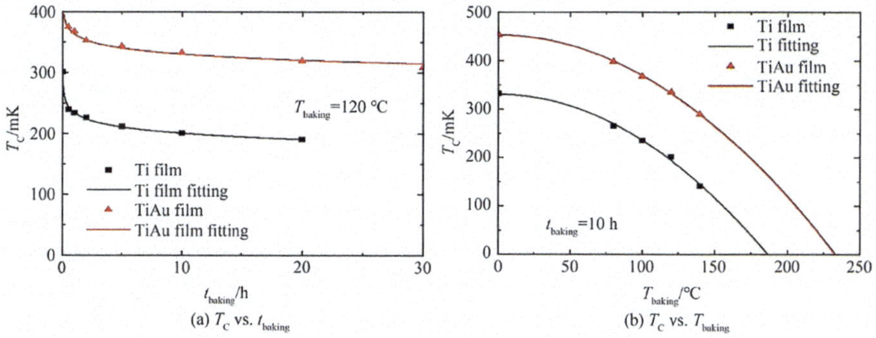


**Fig. 3.12** Curves of the $T_C$ of the Ti membrane and TiAu bilayers with baking time and baking temperature

decrease in $T_C$. Transmission electron microscope observations reveal that the thickness of the $TiO_x$ layer after baking is slightly greater than before, and energy dispersive X-ray spectroscopy indicates that the oxygen peak width increases after baking. This suggests that the surface of the titanium film undergoes oxidation, forming a non-superconducting oxide layer. The oxide layer, which is between 2 and 5 nm thick, has a negligible effect on absorption. Additionally, the normal resistance of both Ti membranes and TiAu bilayers remains largely unchanged. However, $T_C$ decreases with natural aging, and the change in $T_C$ follows a similar pattern to that observed with increased baking time. This study provides a flexible and convenient method for realizing a high-efficiency superconducting TES single-photon detector with photon number resolution, offering a reliable approach to enhance the performance of these detectors.

C. **Heat transfer mechanism and modeling of the superconducting TES single-photon detector**

A superconducting TES single-photon detector with a very small effective area (1 μm × 1 μm) was developed to characterize its electrical and thermal properties. The measured thermal conductivity (G) and estimated effective thermal relaxation time ($\tau_{eff}$) suggest that, in addition to phonon diffusion cooling, electronic diffusion cooling also plays a significant role. This ultra-small area detector exhibits high sensitivity, speed, and low heat capacity, making it suitable for applications requiring photon number resolution, such as astronomical observation and quantum information.

The phase slip coefficient was approximated as a continuously changing parameter to extend the di-fluid model. The α and β parameters for the superconducting TES single-photon detector were successfully extracted (as shown in Fig. 3.13a). The calculated current–voltage and pulse response curves were found to be in complete agreement with the measured results, confirming the effectiveness of the extended model. The main parameters of the superconducting TES single-photon detector for different sizes were characterized, and the variation patterns of these parameters

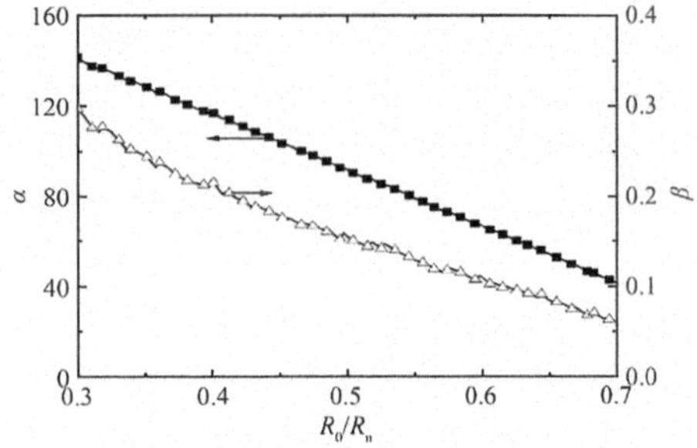

(a) Calculated $\alpha$ and $\beta$ as a function of $R_0/R_n$

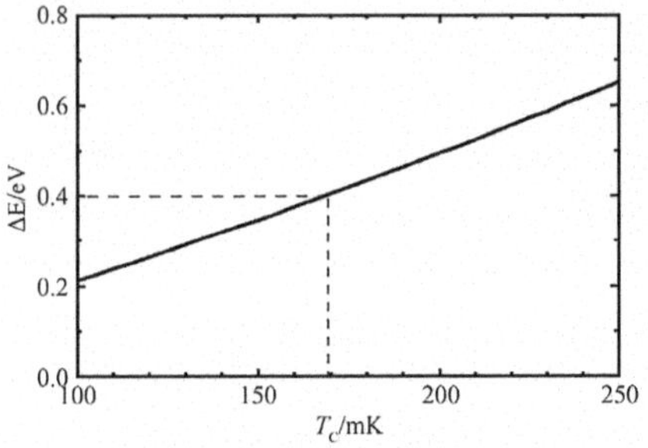

(b) Predicted $\Delta E_{FWHM}$ as a function of $T_C$

**Fig. 3.13** Temperature and current sensitivity coefficient and energy resolution of the superconducting TES single-photon detector

were analyzed. By introducing a noise factor (M), the theoretical calculations for the superconducting TES single-photon detector's performance, considering device size and $T_C$, were obtained. As shown in Fig. 3.13b, a device size of 20 μm × 20 μm and $T_C$ below 170 mK can achieve both high detection efficiency and high energy resolution, providing guidance for further device optimization.

D. **1550 nm wavelength high-performance superconducting TES single photon detector**

A superconducting TES single-photon detector based on titanium film was prepared to evaluate its resistance transition properties and current–voltage characteristics. For a 10 μm × 10 μm device, the thermal conductance (G) was 288 pW/K, with a $T_C$ of 285 mK. The noise equivalent power (NEP), dominated by phonon noise, was calculated as $NEP_{phonon} = (4k_BT^2{}_CG)^{0.5}$, yielding $NEP_{phonon} = 2.8 \times 10^{-17}$ W/$Hz^{0.5}$ and 0.23 eV, which is much lower than the energy of a 1550 nm photon ($E_\gamma$ = 0.8 eV), enabling photon number resolution at 1550 nm.

To achieve high absorption efficiency and detection efficiency, an integrated optical cavity was developed for the superconducting TES detector. This optical cavity consisted of a dielectric mirror, superconducting film, and anti-reflection layer. The refractive index of the superconducting titanium film was studied using an optical ellipsometer, showing a slight increase in the refractive index with wavelength in the near-infrared band. The titanium film formed a 3 nm thick oxide layer on the surface when prepared on a Si substrate. An optical cavity for the 1550 nm band was designed, as shown in Fig. 3.14a, using 8 cycles of $SiO_2$ and 2 cycles of $Ta_2O_5$. The dielectric mirror layers had a thickness of 1/4 wavelength, and the anti-reflection layer's thickness was optimized for efficient absorption at 1550 nm. The absorption efficiency after integrating the optical cavity was measured to be 95%. The final superconducting TES detector with the integrated 1550-nm optical cavity exhibited a thermal conductance (G) of 60 pW/K, an effective thermal relaxation time ($\tau_{eff}$) of 5.9 μs, and an energy detection efficiency of 40%.

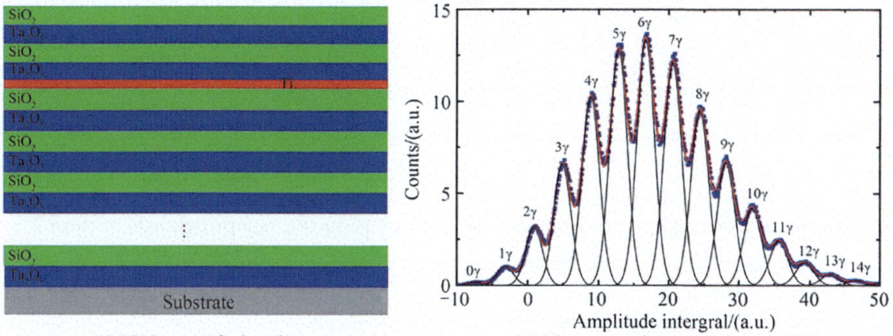


(a) 1550 nm optical cavity (b) Histogram of superconducting TES

**Fig. 3.14** 15 Optical cavity and response height histogram of the superconducting TES single-photon detector

A 1550 nm optical cavity was further optimized to control the thickness of each layer precisely, achieving an absorption efficiency ($\eta_{absorption}$) of over 95%. The titanium film thickness was reduced to 22 nm, and the device size was set to 20 μm × 20 μm, which is significantly larger than the spot size of single-mode fiber, achieving near 100% coupling efficiency. The device's $T_C$ was lowered to 160 mK, enabling it to distinguish 1550 nm photons, though the dark count was high (about kcps). Optical signal transmission efficiency was enhanced by using optical fiber fusion technology, eliminating low-temperature joints on the bare optical fiber ($\eta_{fiber}$).

The spectral ratio (R) of the splitter and attenuator combination was $0.676 \times 10^{-4}$. With the picosecond pulse laser set to a repetition frequency (f) of 20 kHz, the optical attenuator's attenuation value was adjusted, and the average input photon number was calculated to be 6.86. The histogram in Fig. 3.14b shows the superconducting TES single-photon detector's response height, with the average number of photons detected being 6.33. Thus, the system detection efficiency ($\eta_{sys}$) was 92.2%, with an energy resolution of 0.55 eV. The detector can distinguish at least 12 photons at 1550 nm.

The system detection efficiency ($\eta_{sys}$) is determined by $\eta_{fiber}$, $\eta_{absorption}$, fiber coupling efficiency ($\eta_{coupling}$), and the TES device detection efficiency ($\eta_{TES}$), where $\eta_{TES}$ is 100%, $\eta_{absorption}$ is 95%, and $\eta_{fiber}$ is 98%. To further improve $\eta_{sys}$, a more refined optical cavity will be prepared to enhance $\eta_{absorption}$ and reduce the distance between the fiber end surface and the device to improve $\eta_{coupling}$.

In further optimization, a fiber shielding system was added, and the fiber was coiled into a 35 mm diameter at low temperatures to form a long-wave filter to block infrared light. A smaller device (5 μm × 5 μm) was selected, improving the current response rate of the superconducting TES detector to 0.26 eV, with a system detection efficiency of 36% and a dark count of 4 ± 2 cps. The optimized detector can still distinguish at least 12 photons at 1550 nm, achieving a superconducting TES single-photon detector with low dark count.

### E. Visible light band superconducting TES single-photon detector

A superconducting TES single-photon detector with an integrated 850 nm optical cavity and a titanium film thickness of 53 nm was successfully developed for the visible band. The device's $T_C$ is 400 mK, and its normal state resistance (Rn) is 13Ω. Accurate alignment with SM800 single-mode fiber (mode field diameter of 5.6 μm) was achieved using an infrared microscope and a precision 2D mobile platform. A pulsed laser was used to attenuate the number of photons per pulse to less than one photon, and the response waveform was recorded using a digital oscilloscope.

At 300 mK ambient temperature, the device was biased at 0.74 Rn, and response waveforms were recorded for single-, two-, and three-photon cases. As shown in Fig. 3.15 (a), the response height increased with the number of photons, demonstrating the photon number resolution capability of the superconducting TES detector. Using a data acquisition card, the response height was statistically analyzed, yielding a histogram of the response height distribution. The results showed that at 300 mK, with a photon energy of 0.75 eV, the system detection efficiency ($\eta_{sys}$) was 13%, and the detector could resolve at least 10^8 850-nm photons.

The 850 nm band superconducting TES single-photon detector was provided to the Institute of Optics and Laser Metrology at the Chinese Institute of Metrology. It was applied to a single-photon radiation reference measurement device, serving photon precision measurement and optical radiation measurement reference technology research, as shown in Fig. 3.15b.

A highly sensitive superconducting detector, particularly a superconducting TES single-photon detector, plays a crucial role in astronomical research. In the visible and near-infrared bands, it offers significant advantages in detection efficiency, energy resolution, photon number resolution, and dark count. This makes it ideal for studying weak celestial signal sources and fast time-varying sources by maximizing photon

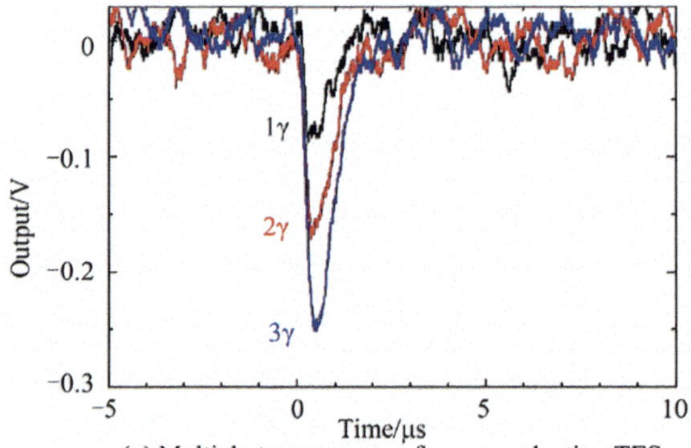


(a) Multiphoton response of superconducting TES

(b) Setup of single-photon emission

**Fig. 3.15** Multi-photon response and single-photon radiation reference measurement device for the superconducting TES single-photon detector information. The characteristics of titanium membranes and the $T_C$ regulation mechanism are thoroughly discussed. The di-fluid model is expanded to extract key parameters of the superconducting TES single-photon detector, and theoretical calculations demonstrate the dependence of $T_C$ and energy resolution. An integrated optical cavity was developed for the titanium film-based superconducting TES single-photon detector, achieving a system detection efficiency greater than 90%. The response time of the detector is on the order of microseconds, meeting the demands of high time-resolved astronomical observation.

(3) **Research on miniaturization of free operation InGaAs/InP single-photon detector**

A. **Design and implementation of the free-running single-photon detector**

The basic circuit diagram of the QCD600 series using NFAD to achieve free-running SPD is shown in Fig. 3.16.

DC voltage $V_{bias}$ provides reverse bias voltage for NFAD. On the bias link, $R_1$ is the avalanche current sampling resistance, $R_L$ is the integrated negative feedback quenching resistance of the NFAD, and $R_2$ is the dead time signal impedance matching resistance of 50 Ω. When an avalanche event occurs, the avalanche current generates a pressure drop on $R_1$ to draw the avalanche signal through the AC coupling capacitor $C_1$. In the NFAD, the avalanche time is very short, on the order of 100 picoseconds, and the initial avalanche signal amplitude is small, around 1 to 2 mV. To boost this, the signal is first amplified by 40 dB using a cascade of low-noise amplifiers (LNA), increasing the amplitude to 100 to 200 mV. It is then digitized

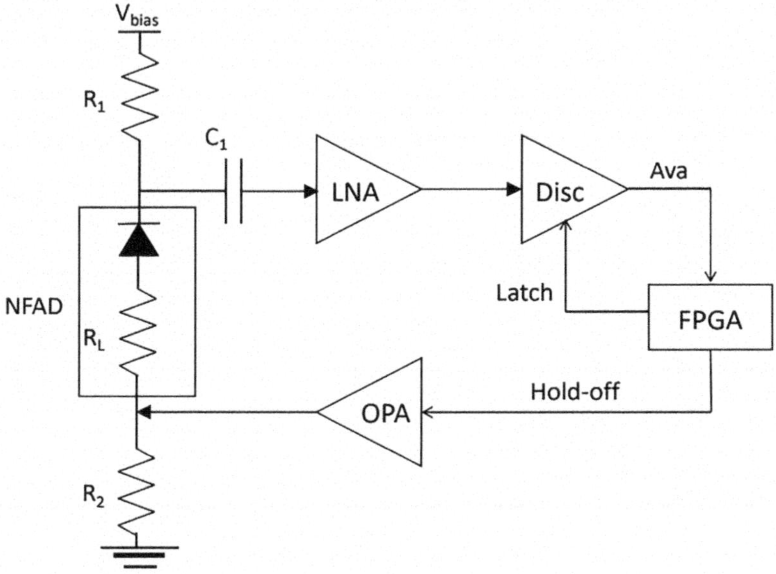


**Fig. 3.16** Schematic diagram of the free-running SPD

by a discriminator and sent to a field programmable gate array (FPGA) for further processing.

The performance of NFAD is significantly influenced by temperature, and they typically require cryogenic cooling to reduce the DCR. However, at lower temperatures, the lifetime of trapped carriers increases, leading to a higher $P_{ap}$. To mitigate this, the FPGA sets an adjustable dead time after each avalanche event to suppress after-pulses, though this comes at the cost of a reduced saturation count rate.

The dead time signal, which is initially at 3.3 V from the FPGA, is increased to above 6 V using an operational amplifier (OPA) to meet the bias requirements of the NFAD. This ensures that the overvoltage is kept below the avalanche breakdown voltage when the overvoltage approaches 5 V. Applying the dead time signal to the NFAD anode produces a sharp differential response, which is latched using the latch function, filtering out any differential signal effects and improving the accuracy of the measurements. The device design also includes bias control, bias current detection, refrigeration temperature control, time-to-digital conversion (TDC), and after-pulse correction algorithms for optimal performance.

(a) **Polarization control and deflection flow detection**

The bias directly determines the performance index of the single-photon detector, and its stability is crucial. Because the avalanche breakdown voltage of NFAD is generally 50–80 V, DC/DC boost chip LT3482 (Linear Technology) is selected, its maximum output voltage can reach 90 V, with current mirror output pin, which is easy to monitor the current bias to implement strong light protection. The principle of bias control and bias current monitoring circuit is shown in Fig. 3.17.

The FPGA controls the output voltage of the 16-bit DAC (Digital to Analog Converter) chip via SPI (Serial Peripheral Interface) chip LT3482 with TPC116S4 (3PEAK Incorporated) bus, At the same time, through the 16-bit ADC (Analog to

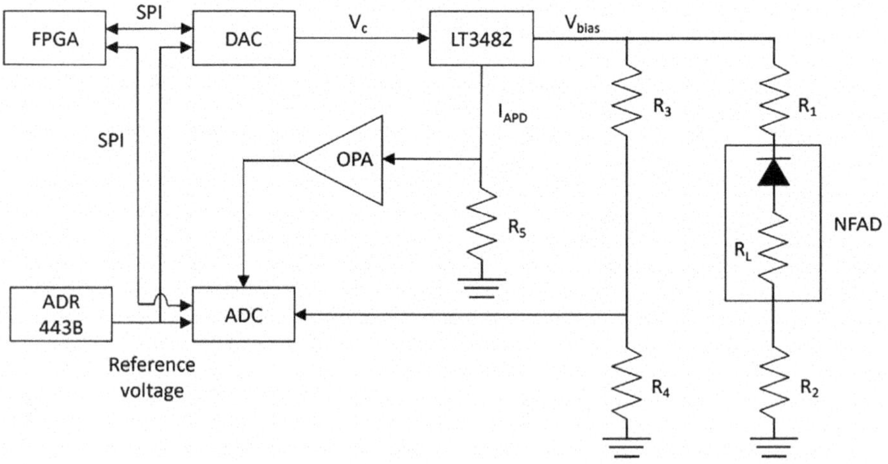


**Fig. 3.17** Schematic diagram of the bias control and bias current detection circuit

Digital Converter) chip AD7682 (ADI) sampling of high-precision low-temperature drift resistance $R_3$ and $R_4$ partial voltage, Biasing stability is achieved through FPGA feedback regulation algorithm using partial pressure monitoring value, The amplitude of bias can be controlled to less than 50 mV with the temperature change. Bicurrent detection is to detect the magnitude of NFAD operating current by detecting the magnitude of the current at the output of the mirror current source of the LT3482. The current $I_{APD}$ is 20% of LT3482 output current. $I_{APD}$ is converted into a voltage signal through resistance R5, and then amplified by OPA by AD7682 for sampling monitoring. The monitoring value minus the partial voltage detection road consumption current value is the working bias of NFAD. FPGA thresholds the bias value to detect whether NFAD receives a strong light signal and reports the abnormality. When the abnormality occurs, the NFAD protection under strong light injection is realized by reducing the output voltage of LT3482.

(b) **Refrigeration temperature control**

Temperature is another decisive factor in the performance index of single-photon detectors. QCD600 Series of NFAD use the TO-46 package with fiber coupling. Through the design of tube clips and refrigeration boxes, it uses a semiconductor refrigerator (Thermo Electric Cooler, TEC) to provide refrigeration. The lowest operational temperature of NFAD can be up to − 50 °C, which can effectively suppress the dark count. The principle of the refrigeration temperature control circuit is shown in Fig. 3.18.

Two partial pressures were differentially amplified by OPA and collected by ADC. The ADC code corresponds to the pt100 temperature. Compared with the current temperature and target temperature, the DAC output voltage is adjusted through the

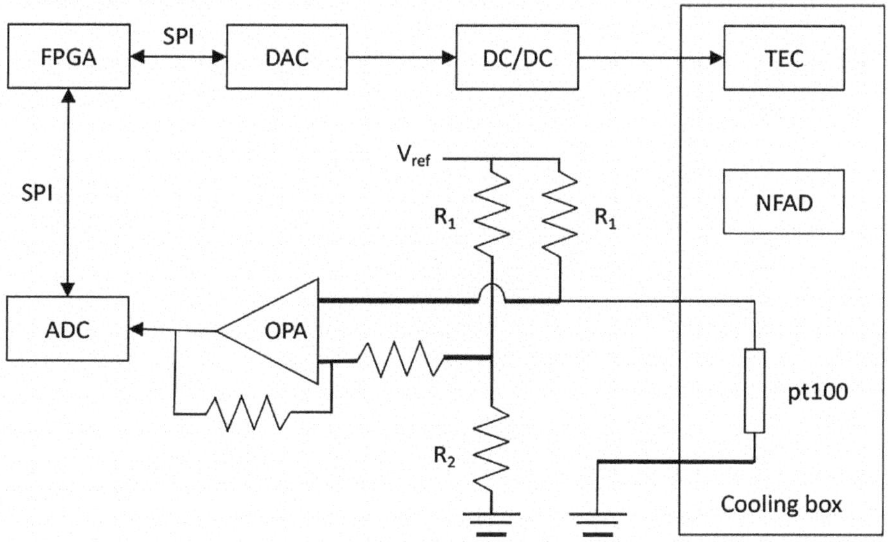


Fig. 3.18 Schematic diagram of the refrigeration temperature control circuit

digital PID algorithm inside FPGA to control the output voltage of the DC / DC power supply, to control the cooling power of TEC. Finally, the cooling temperature of NFAD can be gradually stabilized to the range of the set temperature value ± 0.1 °C, effectively ensuring the stability of the detector performance.

On the other hand, reliable detector products need to adapt to a wider ambient temperature range. When the ambient temperature rises to 50 °C or even higher, due to the limitation of TEC cooling capacity, it cannot maintain the cooling temperature of − 50 °C for a long time, and the performance of the detector will fluctuate. Through the FPGA adaptive algorithm, adjust the refrigeration target temperature, and according to the linear relationship between the avalanche voltage and temperature of the working bias value, under the sacrifice of certain dark count performance can make the detection efficiency remains unchanged and obtain better pulse performance, while indirectly avoid the TEC long working in the limit conditions, extend the service life of TEC and the whole detector.

(c) **TDC function**

In a lidar system, the measurement of photon flight time is a very important link, and the accuracy of time measurement directly determines the distance resolution of the lidar. The integration of the high-precision TDC function in the detector can effectively improve the integration and performance of the lidar system. The QCD600 series detector has an external input trigger signal interface, which is sent to the FPGA as the beginning of the TDC (Start) signal, while inside the detector, the detection signal is directly sent to the FPGA as the end (Stop) signal of TDC. In the FPGA, on the one hand, measuring the coarse time to realize 10 ns coarse time interval and 100 μs range; on the other hand, using the lowest transmission delay of 100 ps, realizing the fine time measurement by delay chain interpolation, and finally realize the time measurement with wide range and high precision. The design principle of fine time measurement based on the carry delay chain is shown in Fig. 3.19.

The Start and stop signals pass through the buffer (buffer) into the negative signal and the positive signal with fixed narrow pulse width, respectively. The start signal is sent into the carry chain, and the stop signal serves as the clock of the latch register. For each stop signal generated, the decoder (Decoder) module encodes the latched data Q 1 to QN as the time information, corresponding to the address bit stored in the memory (RAM), and the time information extraction can be completed by reading out the address data. The detector is designed with a high-speed USB interface, which accumulates the time information of each detection event in RAM, and outputs the time-count rate histogram after accumulation per second through the USB interface, which can realize radar information processing and detector performance self-calibration and other functions with the data processing algorithm.

(d) **After-pulse correction and count rate correction algorithm**

Pap (i) Due to the high concentration of impurities and defects in the material, the InGaAs/InP single-photon detector usually has a large after-pulse effect, requiring increasing the dead time to suppress the after-pulse. In the application of single-photon lidar with high dynamic range requirements, both the after pulse and the

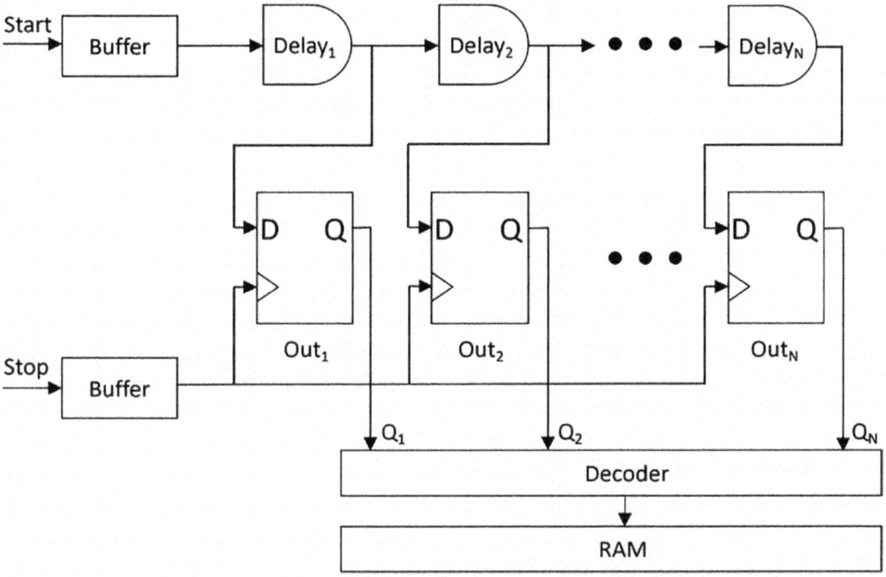


**Fig. 3.19** Block diagram of TDC design principle based on carry-delay chain

dead time will distort the radar echo signal, so the distribution of the original echo light signal should be corrected by the post-processing algorithm. The probability distribution of the after-pulse counts of a single avalanche signal at subsequent times can be calculated by inputting narrow-spike light using the time-count distribution of TDC, where i represents the horizontal axis time bin serial number of the time-count distribution. In the measured raw Lidar echo signal, the probability of the avalanche counting in the $m_{th}$ bin at the nth bin is:

$$P(m, n) = P_{ap}(n - m) \cdot P_{nc}(m, n) \cdot P_{nap}(m, n).$$

The probability that no after-pulse count is generated between $m_{th}$ and $n_{th}$ bin can be calculated by the following formula:

$$P_{nc}(m, n) = \exp\left(-\frac{1}{R}\sum_{k=m}^{n} \text{count}(k)\right)$$

$$P_{\text{nap}}(m, n) = \exp\left(-\sum_{k=1}^{n-m} P_{ap}(k)\right)$$

where R is the total count. For the count rate correction, since the count (m) in the m bin is generated in the condition of no avalanche count in the m bin, where td represents the number of bins corresponding to the dead time, it can be obtained:

$$\text{correct\_ count (m)}=\frac{\text{count(m)}}{1\text{ - p(m)}}$$

where p (m) indicates the probability of avalanche counting in m td bin to m 1 bin:

$$p(m) = \frac{1}{R}\sum_{k=m-t_d}^{m-1} \text{count}(k)$$

QCD600 Series detectors have a built-in single-chip microcomputer (Micro Controller Unit, MCU), using the above algorithm for after-pulse correction and count rate correction of TDC data. FPGA transmits the real-time TDC data from RAM to MCU. MCU corrects the data through a USB interface to realize real-time data post-processing.

(e) **The detector structures**

QCD600 Series detectors include the main control panel, cooling box, radiator, fan, and product shell. The product explosion diagram and appearance are shown in Fig. 3.20. The length and width of the product are 116 mm × 107.5 mm × 80 mm, and the volume is less than 1 $dm^3$, as contrast ID Qube, width and height are 95 mm × 95 mm × 95 mm. The refrigeration box uses optical fiber plug dispensing and parallel sealing and welding technology to ensure air tightness and ensure the reliability of the product.

B. **Test of the detector performance indicators**

Detector performance is calibrated using a narrow-pulse light scheme attenuated to the single-photon level. For weak coherent light sources, the photon number conforms to the Poisson distribution, and the detection efficiency of the SPD is calculated as follows:

$$\eta = -\frac{1}{\mu}\ln\left(1 - \frac{\text{count( coin )}}{f}\right)$$

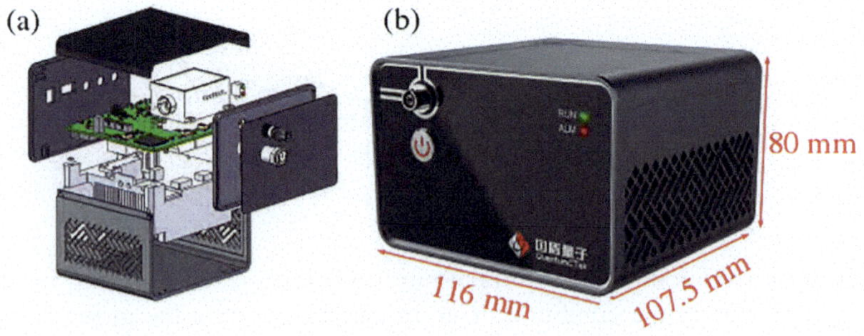


**Fig. 3.20** **a** Product explosion diagram; **b** appearance

**Table 3.1** Detection efficiencies at different wavelengths

| Wavelength/nm | Detection efficiency (%) |
|---|---|
| 1550 | 25 |
| 1310 | 35 |
| 1064 | 35 |

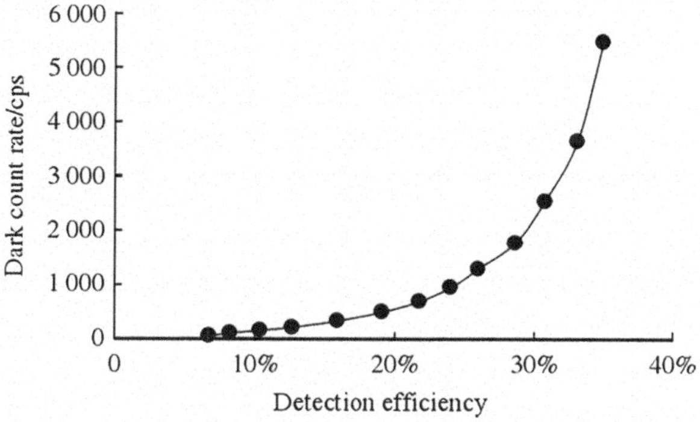


**Fig. 3.21** Detection efficiency—DCR curve

In the actual test, at 1 and 50 kHz, the cooling temperature of NFAD was − 45 °C, and the dead time was set to 5 μs. First, the response wavelength of core NFAD ranges from 900 to 1700 nm for 1550, 1310 and 1064 nm commonly used in radar applications, and the detection efficiency of 1550 nm is 25% as the reference. The test results are shown in Table 3.1.

Second, test the detection efficiency- DCR curve at 1550 nm wavelength, as shown in Fig. 3.21. At 15% detection efficiency, the DCR can to 250 cps; at 25% detection efficiency, the DCR is only 1 kcps.

Finally, the detection efficiency-time jitter curve was tested, as shown in Fig. 3.22. It can be seen that the higher the detection efficiency, the less the time jitter. At the detection efficiency of 35%, the minimum time jitter can be up to 80 ps.

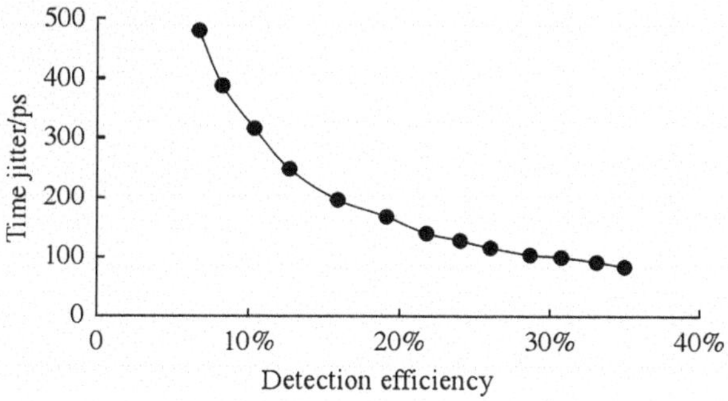


**Fig. 3.22** Detection efficiency-time jitter curve

C. **The application of detectors in radar**

Single-photon detector is used in lidar systems, which can effectively improve the detection distance and resolution. Thanks to the advantages of a multi-mode fiber coupling interface, integrated TDC function and real-time after-pulse and count rate correction algorithm, the application of QCD600 series detectors in 1550 nm atmospheric aerosol lidar greatly improves the integration of radar system and reduces the cost. Figure 3.23 shows the miniaturized lidar product of Guoyao Quantum Radar Technology Co., LTD. Using a QCD600 series SPD. Its single pulse energy is only 75 0 μJ, the farthest detection distance can reach 15 km, the distance resolution is less than 30 m, and can adapt to the working ambient temperature of − 25–40 °C. By contrast, it is difficult to achieve a detection range of more than 10 km.

Figure 3.24 shows the lidar signal collected by the lidar product using a QCD600 series SPD. The data comparison before and after the rear pulse and count rate correction effectively reduces the radar signal distortion caused by the rear pulse and dead time of the detector.

The QCD600 series of miniaturized free-running InGaAs/InP single-photon detector provides a compact SPD solution for the near-infrared band with high efficiency, low noise, low time jitter and real-time data post-processing. In the future, free-running SPD products will develop in the smaller direction of using integrated refrigeration technology and the ultra-low noise direction of using deep refrigeration technology, providing more powerful technical support for lidar, QKD and other applications.

(4) **Development and application of InP-based single-photon detector**

Photoelectric detection in the near-infrared band, particularly at 1550 nm, is a focal point of application research due to several key advantages. This band lies within the communication window, offering high atmospheric transmittance, low optical

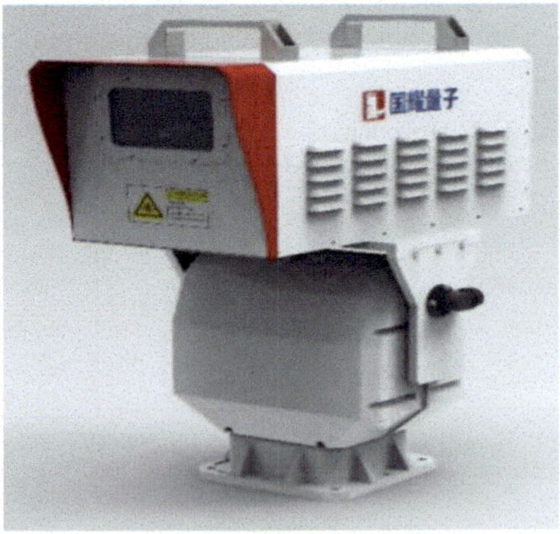

**Fig. 3.23** Lidar products using QCD600 series of single-photon detectors

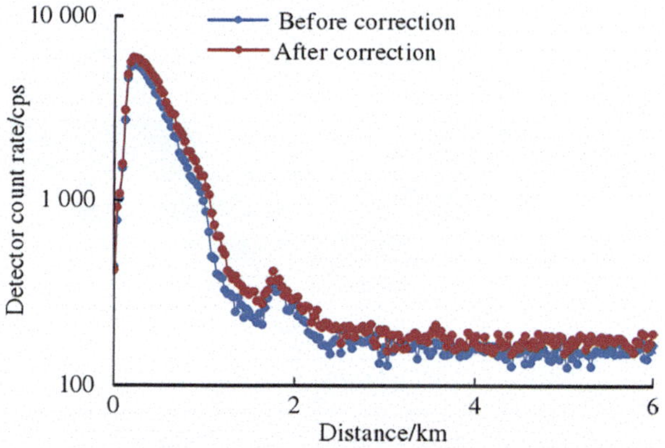


**Fig. 3.24** Comparison of data before and after lidar signal correction

fiber transmission loss, and excellent concealment characteristics. It also experiences minimal interference from sunlight, ensuring clearer signal reception, and is safe for human eyes. As a result, the near-infrared band is widely utilized across military, commercial, and civil industries.

Current devices for effective near-infrared SPD include PMT, superconducting SPDs, QD field-effect tubes, and frequency up-conversion detectors based on Si single-photon detectors. However, each has its drawbacks: PMTs are large and difficult to integrate; superconducting single-photon detectors are complex, costly, and require extremely low 4 K temperatures; and frequency up-conversion systems are complex and unsuitable for large-scale applications. In contrast, the InP-based GM-APD single-photon detector stands out for its high quantum efficiency, compact size, ease of integration into arrays, low power consumption, and low cost, making it highly valuable for near-infrared band SPD.

A. **Recent development of InP/InGaAs single-photon detectors for SAGCM structures**

The SAGCM structure [22, 23] consists of an absorption multiplier layer, a slow change layer, and a charge layer, which is the foundational design for most InP-based APDs. This structure generates high fields in the InP multiplication region to enable collision ionization for avalanche gain, while maintaining lower fields in the $In_{0.53}Ga_{0.47}As$ absorption layer to prevent tunneling effects. The charge layer between the absorption and multiplication layers, known as the field control layer, allows for flexible adjustment of the internal electric field distribution. The indium gallium arsenide phosphide (InGaAsP) gradient layer minimizes hole accumulation at the InGaAs/InP heterojunction (Fig. 3.25).

The InGaAs/InP SPAD is designed with a structure that includes an $In_{0.53}Ga_{0.47}As$ absorption layer, which determines the device's long-wave spectral response, and an InP material for short-wave response. The InGaAs absorption layer has a band gap of 0.75 eV, corresponding to a 1.67 μm cutoff wavelength. Photons with wavelengths shorter than 0.92 μm are absorbed by InP. The device structure utilizes two concentric Zn diffusion areas, with the PN junction buried deeper at the center of the diode. This

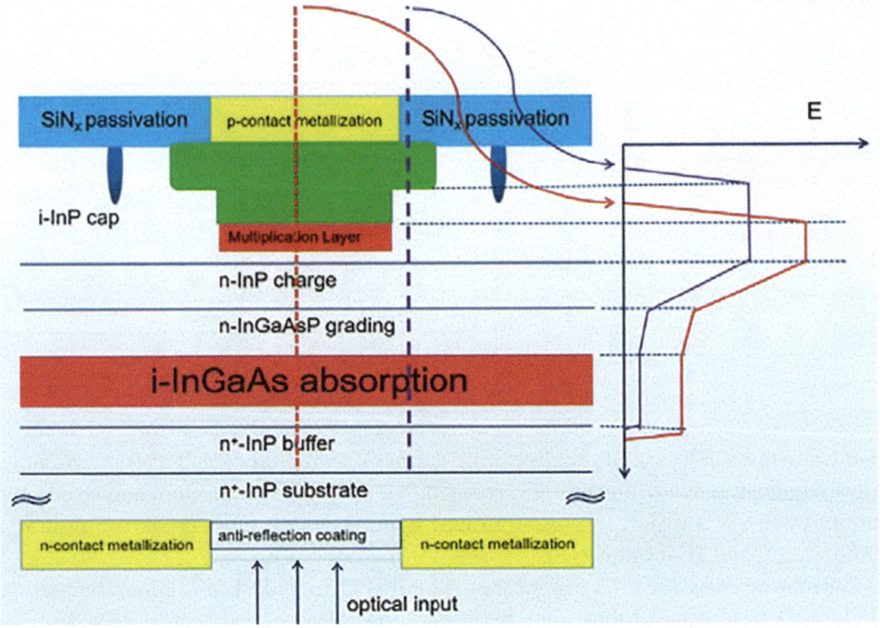


**Fig. 3.25** Geometric avalanche diode structure of the diffusion junction plane

design ensures avalanche breakdown only occurs in the central region of the device. The device's multiplier layer thickness and electric field distribution are controlled by Zn diffusion, which affects the PDE and DCR.

Key performance indicators of the SPAD include detection efficiency, DCR, $P_{ap}$, and time jitter. Detection efficiency refers to the probability of a photon triggering a detection event, while DCR refers to detection events without incident photons. The $P_{ap}$ arises from defects in the multiplier layer, and time jitter measures fluctuations in the photon detection time. To improve performance, self-quenching and negative feedback methods are used to reduce after-pulse effects and avalanche gain.

Research on InGaAs/InP SPADs has been conducted by institutions such as MIT, PLI, and Chongqing Institute of Optoelectronic Technology, achieving significant improvements. PLI's 2007 report on a back-incident InGaAs/InP SPAD with a planar structure showed 45% PDE at 200 K, with DCR below 3 kHz and time jitter of 30 ps. Further improvements have raised the working temperature to 233–243 K, with PDE between 20 and 45% and DCR below 3 kHz.

In 2020, the University of Science and Technology of China reported a high-frequency gated InGaAs/InP SPAD with a detection efficiency of 60% at 1550 nm. This was achieved by optimizing the dielectric-metal reflection layer and developing a weak avalanche-integrated reading circuit (MIRC) to reduce parasitic capacitance and minimize after-pulse effects. The optimized device had a PDE of 60%, DCR of 340 kHz, and $P_{ap}$ of 14.8%. Further optimization allowed for a compromise between PDE and DCR, achieving 3 kHz DCR and 5.5% Pap, with a PDE of 5.5%, as shown

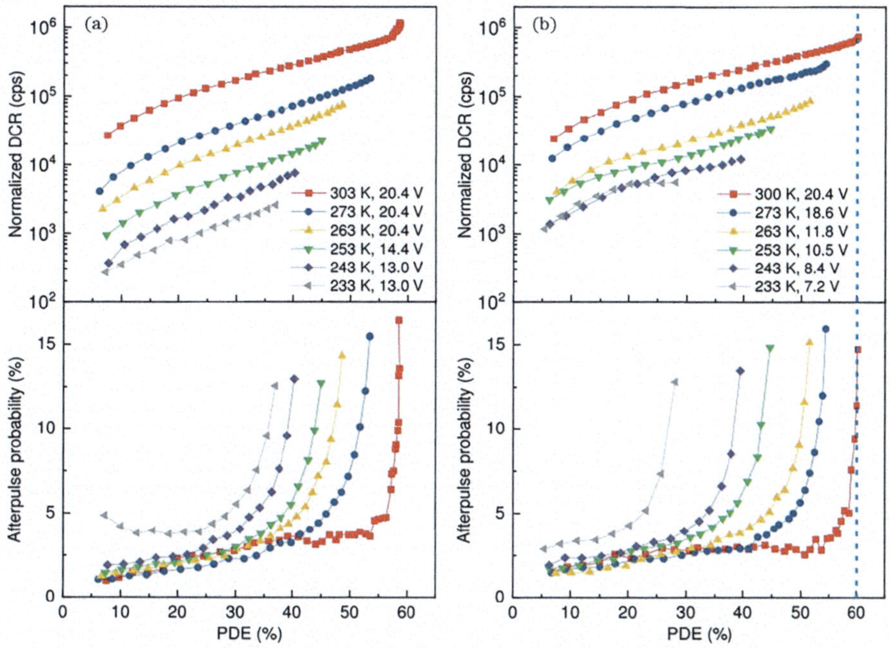


**Fig. 3.26** Relationship between gate height, temperature and normalized DCR, and after-pulse probability and PDE. **a** SPAD No. 1; **b** SPAD No. 2

in Fig. 3.26. This represents the highest reported detection efficiency for a 1550 nm InGaAs/InP SPAD.

B. **New materials with a low factor k**

The gain of SPADs is generated by collision ionization, and each carrier collision ionization is random, resulting in the output gain being random, the noise generated by the excess noise is inherent in SPSDs, an important factor that determines the device SNR mechanism, its size is represented by the excess noise factor. McIntyre shows that the excess noise factor is associated with the device gain and the k factor of the device, and the k factor is defined as the ratio of the hole ionization coefficient to the electron ionization coefficient. In the wide avalanche region, under the condition of electron injection, the excess noise factor is expressed as

$$F = kM + \left(2 - \frac{1}{M}\right)(1 - k)$$

where: M is the gain of the device. By selecting different gain values, the calculated excess noise results are shown in Fig. 3.27. If $\alpha$ and $\beta$ are close, where material k is near 1, the excess noise factor also increases rapidly as the gain increases. The ideal multiplier material k should be close to 0, so the excess noise is not greater

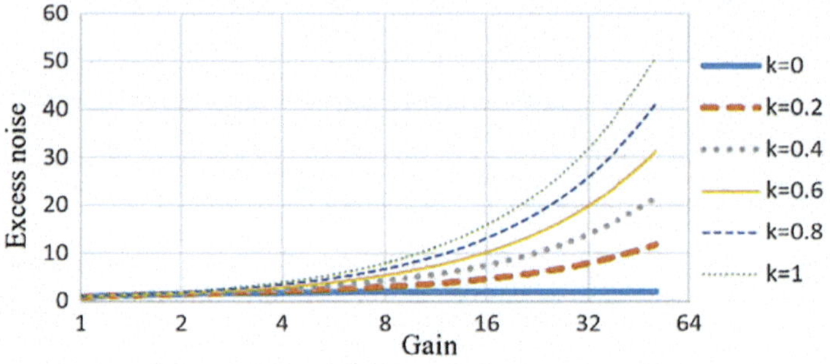


**Fig. 3.27** The excess noise factor and the device gain relationship

than 2. At this time, the multiplier gain is controlled by a photo borne carrier, and the excess noise will be reduced. The smaller k factor represents a lower excess noise factor, so choosing multiplier materials with low K values is very important for high-performance avalanche diodes.

At low temperatures, k = 0 of HgCdTe is the best result, with k factor ranging from 0.3 to 0.5 for In P and k factor of AlInAs ranging from 0.2 to 0.4, and recent studies show that AlInAsSb material systems have high-temperature stability and very low excess noise (k = 0.01–0.05). Shows that AlAsSb has a k factor as low as 0.005, close to the theoretical minimum, Yi et al. Yuan et al. demonstrated that AlInAsSb has a small k factor and excess noise close to Si. Figure 3.28 shows the low noise characteristics of 2 μm wavelength AlInAsSb/GaSb APD and material system. The red symbol is the test result of the excess noise factor of the device as a function of M. It also shows the excess noise factor curve from 0 to 0.1 calculated using the local field model. According to the comparison of the test results and the theoretical calculations, we can see that the designed A and B devices, k 0.01 ± 0.004, both show extremely low excess noise, which is consistent with the previously measured AlInAsSb devices. AlInAsSb Material system of low noise characteristics, temperature stability, and band gap tunability make it in the near-infrared and infrared spectrum of high sensitivity detector candidate, compared with other materials used in the infrared system, its dependence on low-temperature cooling is less, can work at a higher working temperature, this is very ideal for laser radar, night vision, and thermal imaging applications.

C. **Multi-gain coupled amplification single-photon detector**

In a proposal by Rahman et al. from the University of California, San Diego, the MAGIC (Multiple Amplification and Gain Control) single-photon detector was developed, featuring multiple amplification and gain control mechanisms. The device uses both avalanche multiplication and bipolar gain to amplify signals. These mechanisms are coupled in a positive feedback loop, allowing the gain to be controlled at a moderate level, while the device operates just below the breakdown voltage. This coupling allows the theoretical computational detection efficiency to reach up to 98%.

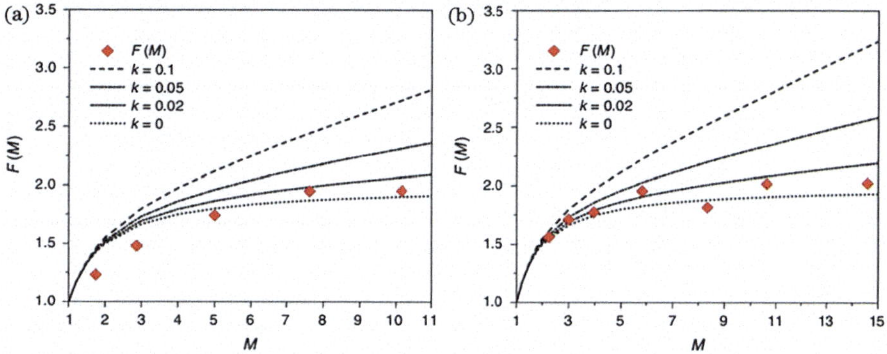


**Fig. 3.28** Measurement curves of the excess noise factor under 2 μm illumination at room temperature. **a** Device A; **b** device B

As shown in Fig. 3.29, the device's principle is equivalent to the coupling of a PN diode with an n-p-n bipolar transistor, where two heterojunctions at the base and emitter form barriers that block electrons and holes. When an incident photon creates an electron–hole pair in the absorption layer, the hole and electron accumulate at the barrier of the heterojunction. This increases the forward bias across the emitter and base, leading to higher bipolar gain. Simultaneously, the increased bias reduces the voltage drop across the multiplier layer, thereby decreasing avalanche gain. When the avalanche gain decreases sufficiently, the base current drops, reducing bipolar gain, and the output current is quenched.

Monte Carlo simulations show that the MAGIC detector's gain distribution is narrower than that of passive quenched APDs, as shown in Fig. 3.30a. This indicates that the randomness of the device's gain is significantly reduced, improving noise performance. Additionally, the MAGIC detector exhibits a larger dynamic range, offering a high resolution from single photon to 25 photons, as shown in Fig. 3.30b. The output current is linearly proportional to the number of input photons, unlike self-quenched and passive quenched APDs, where output current does not correlate with the number of incident photons. The simulated single PDE of the MAGIC detector is up to 98%, demonstrating the potential for SPD in non-Geiger mode devices.

D. **Self-quenched single-photon detector**

The avalanche effect in SPADs creates a positive feedback loop that, once initiated, continues until stopped. To prevent thermal breakdown, negative feedback is required to reduce the bias voltage below the breakdown threshold and halt the avalanche. When integrating SPAD arrays, each SPAD requires its own quenching circuit, increasing process complexity and cost. To address this, a built-in negative feedback SPAD was developed, simplifying the quenching process and enabling easier integration. This design can also operate in free-running mode.

In 2007, a team led by Professor Yu-Hwa Lo at the University of California, San Diego, introduced a self-quenching and self-recovery SPAD based on negative feedback avalanche (ANF). This effect is achieved by placing a minority carrier barrier

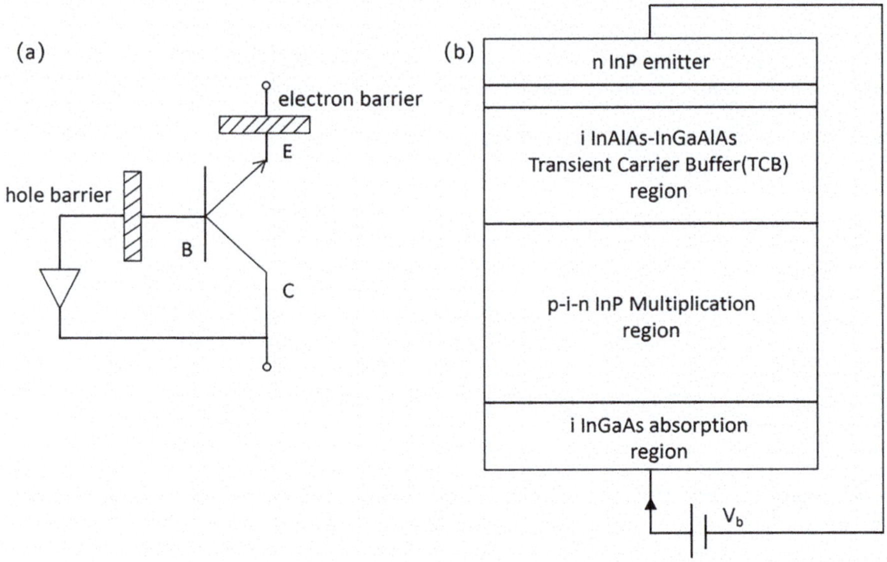


**Fig. 3.29** The MAGIC detector. **a** The equivalent circuit diagram of the MAGIC detector, and **b** the MAGIC detector design using the InGaAs/InP heterostructure

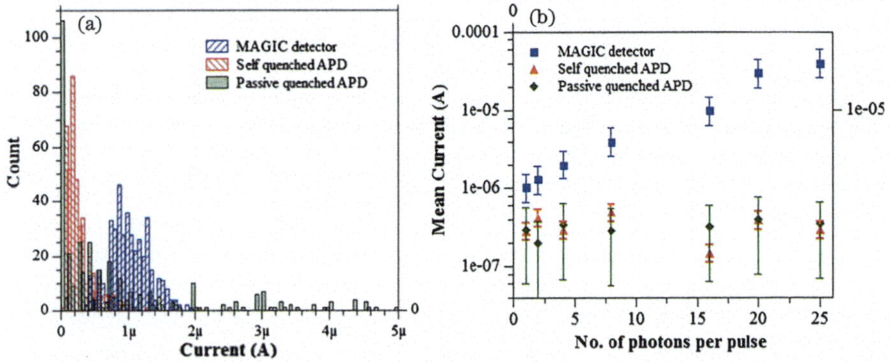


**Fig. 3.30** Results of the Monte Carlo simulations. **a** Monte Carlo simulates the single-photon current response distribution of different devices; **b** the average current amplitude and fluctuation of different photon numbers

near the multiplier region, causing carriers from collision ionization to accumulate at the barrier. This accumulation reduces the electric field and collision ionization, leading to self-quenching without an external circuit. Once the accumulated holes escape the barrier due to thermal excitation or conduction, the device recovers and resumes normal operation.

The self-quenching and self-recovery SPAD [24, 25] reduces gain fluctuations and excess noise, which are common in conventional SPADs. The barrier regulates the carrier concentration, ensuring a uniform quenching process. The device, tested at 160 K, achieved a PDE of 12%, a DCR of 3.3 MHz, and a self-recovery time of 60 ns, as shown in Fig. 3.31.

The barrier in a self-quenching SPAD can be introduced through various methods, such as adding an oxide layer, a broadband gap resistance layer, or using a hetero-junction barrier. As shown in Fig. 3.32, the NFAD, proposed by Princeton Lightwave Jiang, integrates a film resistance within the SPAD structure. This film resistance acts like a passive quenching circuit, with a quenching resistance ($R_s$) placed in series. When the device is operational, the current passing through the integrated quenching resistance forms a voltage drop, reducing the voltage on the SPAD to below the breakdown threshold, thereby achieving avalanche quenching.

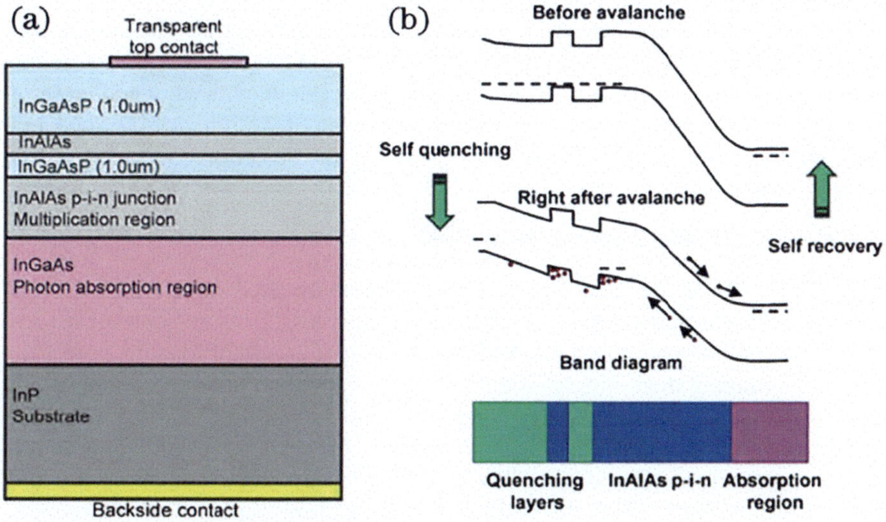


**Fig. 3.31** Structure and working principle of the self-quenched and self-recovery detectors. **a** The structure; **b** the operating principle

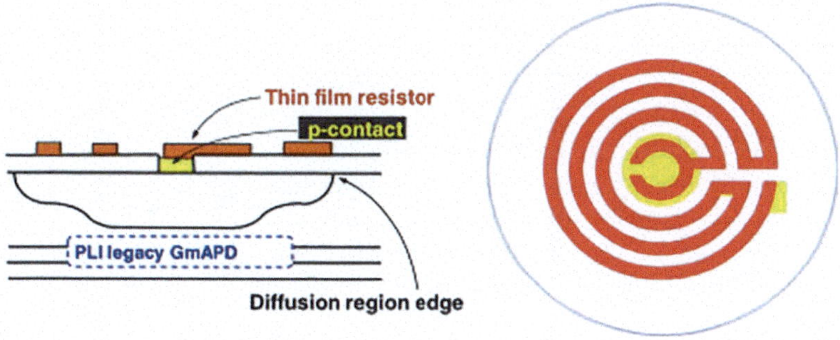


**Fig. 3.32** NFAD top integration resistance

By integrating the quenching resistance, the SPAD's quenching circuit is simplified, parasitic capacitance is reduced, and the after-pulse effect is suppressed. This design also facilitates the integration of SPAD arrays. Moreover, the number of avalanche charges can be controlled by adjusting the bias and temperature, maintaining a controlled charge range of about $10^5$ to $10^6$, which reduces excess noise and enhances the device's performance.

A NFADs offer a significant advantage in their simplicity, as they complete the entire Geiger mode cycle—avalanche, quenching, and charging—under a DC bias. However, their main challenge is the $P_{ap}$, which arises because there is no retention time between quenching and charging to allow trapped carriers to release. Despite this, NFADs can achieve up to 10% PDE, which is higher than SPADs with a controlled hold time. This makes them suitable for certain applications, such as free-space optical communication.

Figure 3.33 illustrates two detectors developed by Itzler et al. The square data represents E3G3, with a quenching resistance of 741 kΩ and a photosensitive area diameter of 32 μm, while the triangle data represents E2G6, with a quenching resistance of 992 kΩ and a photosensitive area diameter of 22 μm. As the detection efficiency increases, the $P_{ap}$ also increases, causing the detection efficiency to drop below 5% when the $P_{ap}$ reaches 10%.

### E. **Geiger mode single-photon focal plane array**

InGaAs/InP SPADs are ideal for detecting commonly used laser bands such as 1550 nm, 1310 nm, and 1064 nm. With the mature process for large-scale array focal planes based on InP materials, InGaAs/InP SPADs offer excellent advantages in array integration, making them highly suitable for applications in lidar, 3D imaging, and more. The InGaAs/InP SPADs array employs a time-counting Si CMOS quenched read-out circuit and connects the detector array to the readout circuit using indium column inversion interconnection. The developed array demonstrates outstanding pixel uniformity, showcasing the advanced manufacturing capabilities of large-area materials and devices.

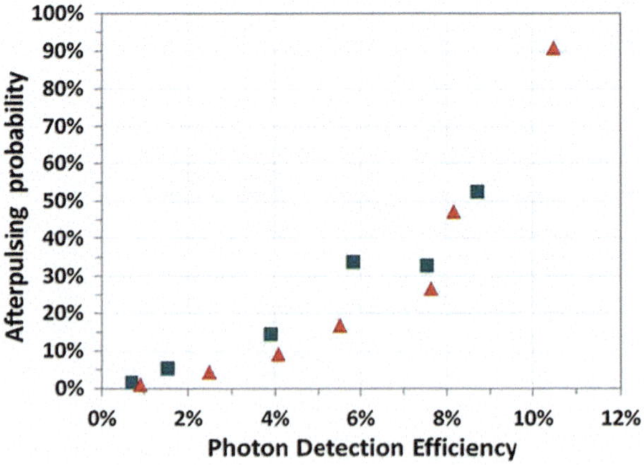


**Fig. 3.33** Relationship between pulse probability and PDE after self-quenching SPADs

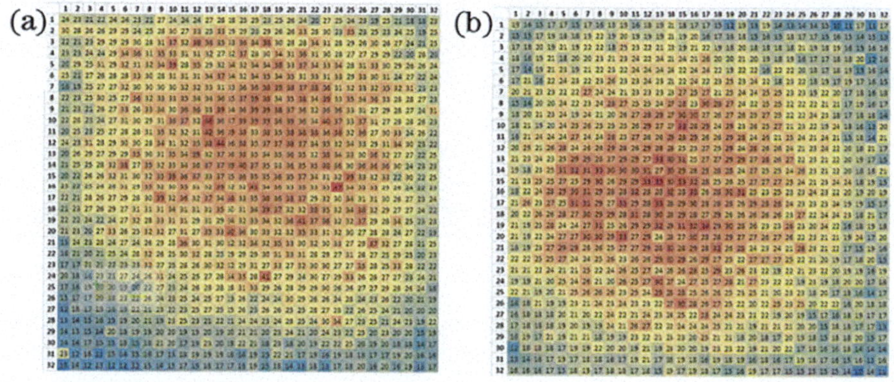


**Fig. 3.34** InGaAs performance of 1.55 μm (InP) SPADs 32 × 32 focal plane array for all 1024 pixels. **a** All pixel DCR units are kHz and less than 50 kHz; **b** an average pixel PDE of 22%, considering all optical losses associated with the microlens array and other insertion losses

Figure 3.34 shows the DCR and PDE test data for a 32 × 32 focal plane array (FPA) with 10 μm pixel spacing at 3.25 V over-bias, operating at a temperature of 253 K. In Fig. 3.34a, the DCR test shows that when the operating temperature is reduced to − 20 °C, the DCR for all 1024 pixels is below 50 kHz, with an average of 28 kHz and a standard deviation of 6.5 kHz. In Fig. 3.34b, the PDE performance shows an average PDE of 22.2% with a standard deviation of 4.6%. These relatively low PDE and DCR values are attributed to the variations in effective over-bias voltage ($V_b$), caused by differences in growth parameters and epitaxial material processes.

For focal plane arrays, uniformity and consistency in pixel response are crucial to avoid spatial noise and maintain high performance. In particular, when pixel spacing decreases, avalanche carriers can couple to adjacent pixels, causing correlated dark counts, referred to as crosstalk. As external bias increases, both detection efficiency and crosstalk rise, which can limit device performance. To reduce crosstalk, strategies include minimizing avalanche charge flow and incorporating physical isolation or absorption structures around the pixel photosensitive zone. One solution is reducing array spacing from 100 to 50 μm, combined with microlens arrays (MLA) and built-in spectral filters, as seen in Fig. 3.35b. Using back-incident techniques, a focal plane with a peak PDE of 45% was achieved for a detection wavelength of 1.55 μm, with dark counts under 20 kHz at 240 K, a dead time of less than 6 μs, and a breakdown voltage difference of about 1 V. The GM-APD array developed by MIT has been applied in flash lidar, photon counting coherent lidar, and communication.

In 2016, Calandri et al. of Cova Group in Italy reported a method to reduce crosstalk by focusing ion beam etching grooves, as shown in Fig. 3.36a, b. By forming the physical isolation between the two pixels and sharing the electric potential with the cathode, the electrical crosstalk of the adjacent pixels can be completely eliminated, but the optical crosstalk introduced by the reflected light between the adjacent pixels cannot be effectively eliminated, as shown in Fig. 3.37.

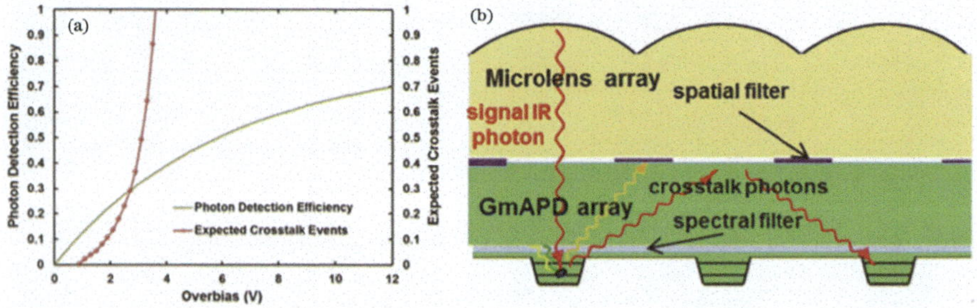


**Fig. 3.35** Schematic representation of the relevant parameters and cross section of GM-APD. **a** The relationship between PDE and crosstalk of GM-APD; **b** the cross-section of InPGM-APD array bound to MLA

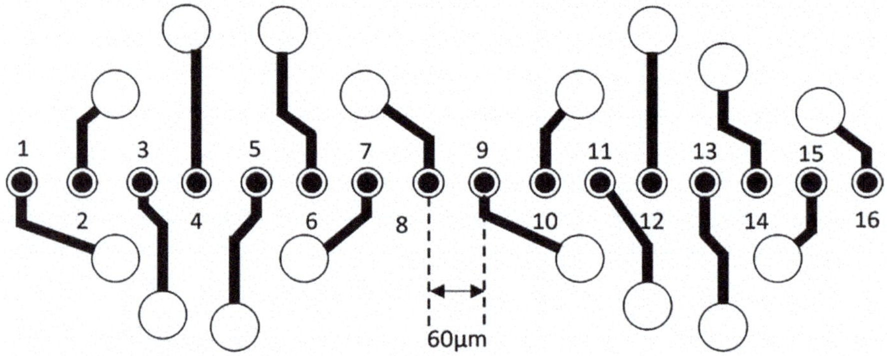


**Fig. 3.36** InGaAs/InP SPAD line array

### F. An InP-based SPD with an extended wavelength

In recent years, expanding the operational wavelength of lidar systems from the conventional 1.55 μm to the 2 μm window has gained significant attention due to growing concerns over human eye safety and the saturation of traditional telecommunication frequency bands. This shift has become crucial for imaging and lidar applications. SPAD devices capable of detecting wavelengths beyond 2 μm, such as InGaAs, HgCdTe, InAsAlSb, and InGaAs/GaAsSb superlattices, are being explored.

For InGaAs devices to respond to wavelengths over 2 μm, the indium content must be increased to nearly 80%, which results in lattice mismatch with InP materials, causing increased material defects and significantly higher dark current density compared to standard $In_{0.53}Ga_{0.47}As$ devices. Additionally, "extended wavelength" InGaAs devices have lower quantum efficiency (QE) of 50% to 60%. HgCdTe material, known for its adjustable bandgap (0–1.5 eV), is an excellent absorber for infrared wavelengths, particularly in the 1–3 μm range, and can achieve a high QE (> 70% at room temperature). However, the high cost of cadmium telluride substrates,

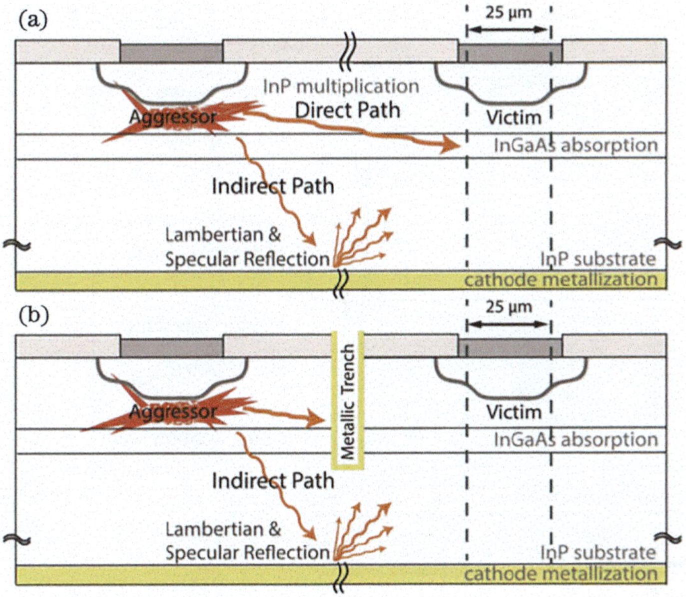


**Fig. 3.37** Schematic diagram of crosstalk between two pixels. **a** Have no metal grooves; **b** have no metal grooves

manufacturing challenges, and low yield hinder its commercial use. Group III-V-based AlInAsSb and InGaAs/GaAsSb superlattice APDs have emerged as promising candidates for extended wavelength single-photon detectors due to their excellent performance and suitable detection range.

(a) **The AlInAsSb SAGCM structure, the APD**

In 2016, Ren et al. demonstrated an AlInAsSb SAGCM APD design for 1.55 μm applications, using a GaSb substrate and adjusting the Al component in AlInAsSb to modify the response wavelength. The design incorporated 40% aluminum in the absorption material and 70% aluminum in the doubling area. This material structure, shown in Fig. 3.38, achieved an excess noise factor (k) of 0.01, which is comparable to or lower than that of Si detectors. The device demonstrated up to 50 normalized external quantum efficiency (EQE) under a 38 V bias, with the optical cut-off wavelength exceeding 1.6 μm. To extend the wavelength further, a thicker $Al_{0.4}In_{0.6}As_{0.6}Sb_{0.4}$ absorption layer and an anti-reflective coating on the top surface can be used, as shown in the test results depicted in Fig. 3.39.

In May 2020, electrical and computer engineers from the University of Virginia and the University of Texas-Austin reported on 2 μm wavelength AlInAsSb/GaSb APDs. The EQE of these devices was approximately 20% without enhanced 2 μm illumination, as shown in Fig. 3.40. The measurements were taken from a 200 μm diameter chip. The Graph A and B chips represent different Al components in the AlInAsSb charge layer: chip A has an Al component of 0.5, while chip B varies

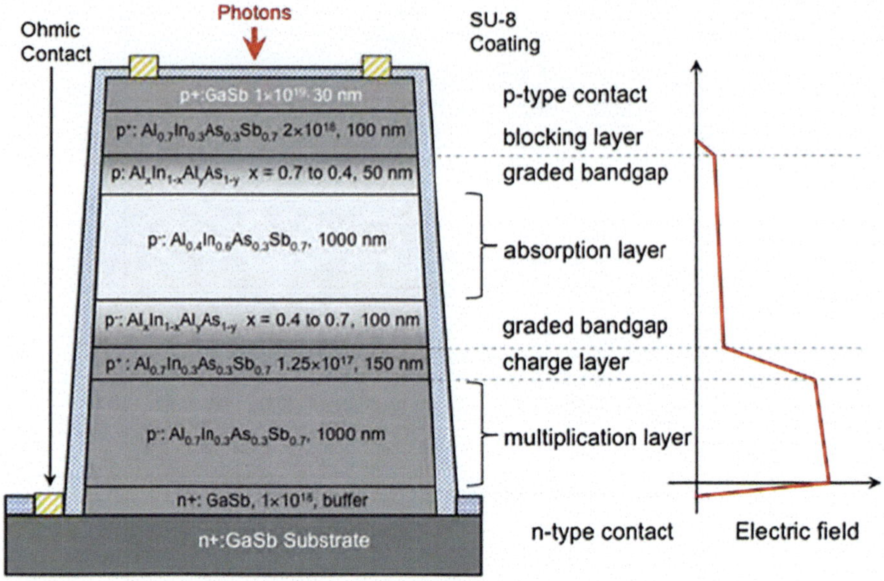


**Fig. 3.38** $Al_xIn_{1-x}As_ySb_{1-y}$SAGCM heterostructure and electric field profile

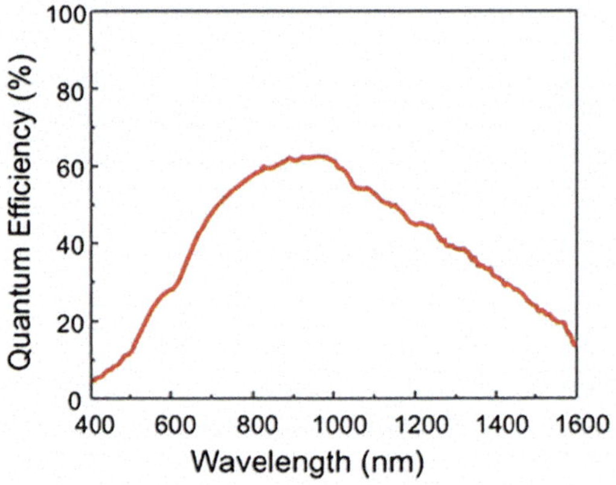


**Fig. 3.39** External quantum efficiency of 150 μm diameter $Al_xIn_{1-x}As_ySb_{1-y}$SAGCMAPD at a temperature of 300 K

between 0.3 and 0.7. Under 2 μm illumination at room temperature, with an 80 μm diameter device, the reverse breakdown voltage was approximately 42 V, the dark current was around $5 \times 10^{-5}$ A, and the gain exceeded 50x, as shown in Fig. 3.41.

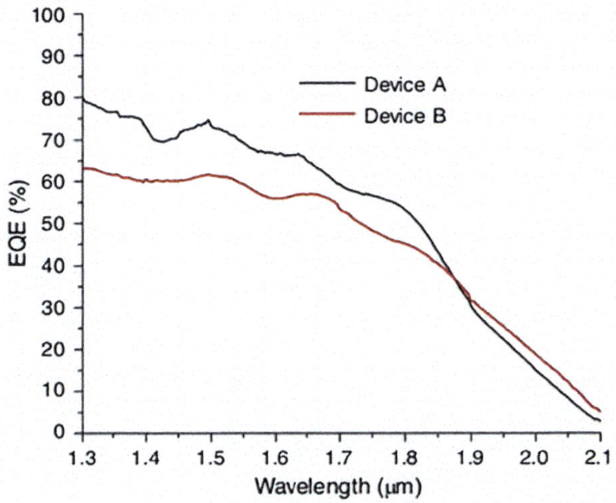


**Fig. 3.40** For devices A and B, EQE is a function of the wavelength at 27 V and 23 V reverse bias, respectively

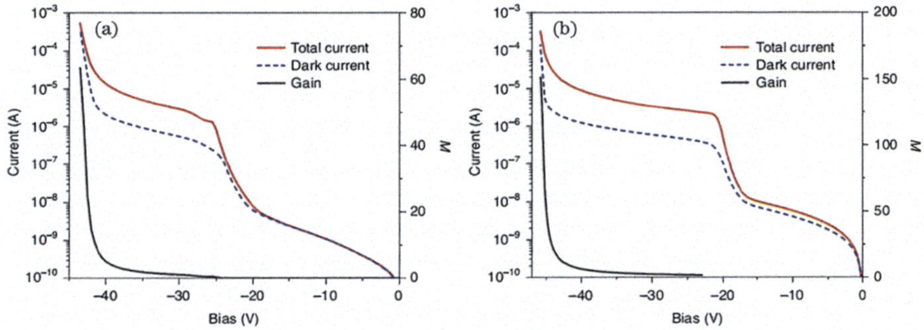


**Fig. 3.41** Current voltage and gain. **a** Device A; **b** device B

### (b) InGaAs/GaAsSb superlattice APD

In 2006 Sidhu et al. reported an InP-based long-wavelength APD that uses a lattice-matched $Ga_{0.47}In_{0.53}As/GaAs_{0.51}Sb_{0.49}$ II-type quantum well for absorption (structure shown in Fig. 3.42). For the 44 μm diameter device, the room temperature dark current at breakdown (37 V) is 130 nA. The dark current decreases rapidly with the temperature, and at 225 K, the dark current at 90% breakdown is about 10 nA, corresponding to current density of 0.66 $mA/cm^2$.

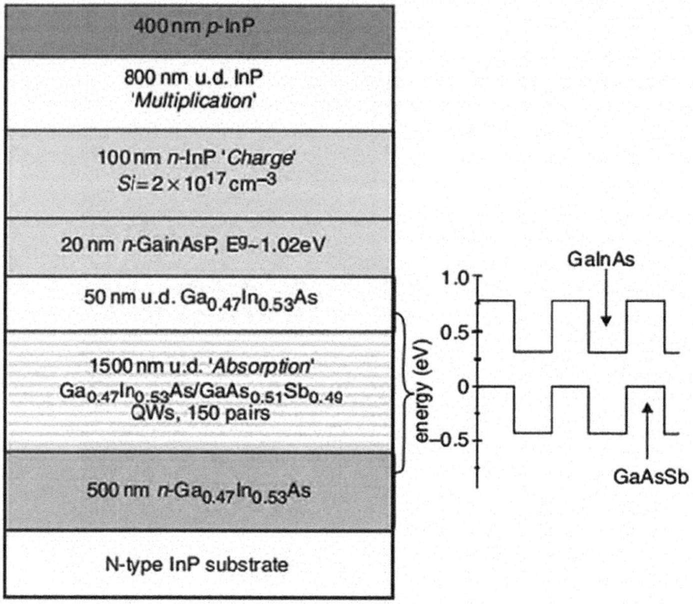


**Fig. 3.42** Schematic diagram and band arrangement of $Ga_{0.47}In_{0.53}As/GaAs_{0.51}Sb_{0.49}$

Recent advancements in InP/InGaAs single-photon detectors have led to significant improvements, making them suitable for applications in quantum communication, lidar, and autonomous driving. Innovations include the development of low K-factor materials to reduce noise, multi-gain amplification mechanisms for sub-Geiger mode detection, and self-quenched single-photon detectors with integrated negative feedback to reduce after-pulse effects. Additionally, extended wavelength detectors using AlInAsSb and InGaAs/GaAsSb superlattices have expanded the detection range. These advancements, combined with improved quenching circuits and larger Geiger-mode SPAD arrays, promise to enhance performance and broaden the application of InGaAs/InP-based detectors.

(5) **Terahertz single-photon detector research**

A. **Performance parameters and testing setup of the THz single-photon detector**

(a) **Basic principles of terahertz single-photon detectors**

A terahertz (THz) single-photon detector is a device that converts THz radiation into electrical signals, enabling the detection of extremely low-energy photons. Two typical types of these detectors are QD detectors (QDD) and Quantum Cooper pair Detectors (QCD), both utilizing "artificial atoms" as their core components. In QDD, the "artificial atom" is a QD, whereas in QCD, it is a single Cooper pair box (SCB). These "artificial atoms" can only accommodate a small number of electrons or Cooper pairs, making them highly sensitive to the tunneling of single electrons or quasiparticles. In both detectors, when photons are absorbed, QDD relies on the energy level transition within the QD, while QCD involves Cooper pairs, where the photon energy exceeds the superconducting energy gap, triggering changes in the system. These mechanisms allow for highly sensitive THz photon detection.

(b) **Performance indicators**

The detection efficiency ($\eta_{tot}$) of a SPD is an important index, which is generally equal to the product of the coupling efficiency ($\eta_c$), absorption efficiency ($\eta_a$) and quantum efficiency ($\eta_q$)

$$\eta_{\text{tot}} = \eta_c \eta_a \eta_q$$

$\eta_c$ refers to the efficiency of the photon coupling from the light source to the detector front, usually related to the area of the detector light sensitive area and alignment, $\eta_a$ refers to the efficiency of the photon absorption by the detector coupling structure, it depends on the design of the coupling structure, $\eta_q$ refers to the photon absorption output response efficiency, usually related to the incident light frequency and the bias conditions of the detector.

DCR is another important indicator of a single-photon detector, which represents the counting rate of the detector in the absence of a photon incident. It is related to factors such as the intrinsic properties of the materials used by the detector and the device bias conditions. The dark count is a misjudgment of a SPD, so the lower the DCR, the better for the SPD.

NEP represents the corresponding incident light power when the effective value of the detector output signal voltage is equal to the noise root mean square voltage value, which represents the minimum power detected by the detector. The lower the NEP, the better the detector performance. For a SPD, NEP and the incident photon frequency (v), DCR and η q satisfy the following equation:

$$\text{NEP} = \frac{hv\sqrt{2\text{DCR}}}{\eta_q}$$

where h is the Planck constant. For a THz SPD, if the photon frequency is 10 THz, NEP is about 110 and η q is $90\%^{-20}$W/Hz$^{1/2}$. This shows that to achieve THz single photon detection, the detector NEP needs to reach $10^{-20}$W/Hz$^{1/2}$ order.

The dynamic range is an indicator to characterize the working range of a SPD. It is the ratio of the maximum value to the minimum value of the light signal that the detector can respond to. The larger the dynamic range, the more abundant the hierarchy that the detector can reflect, which is of great significance in imaging applications. The recovery time also called the dead zone time, is the interval time that a SPD can no longer work after realizing a single photon detection, which determines the maximum count rate of the detector.

(c) **Test the system**

A typical test system for THz single-photon detectors is shown in Fig. 3.43. The THz photons are emitted from a tunable weak THz radiation source, after a series of filters, focused by the lens to the probe. There are two THz sources, narrow band and broadband radiation sources, which usually adopt high-mobility semiconductor materials, such as GaAs/AlGaAs heterojunction and n-type InSb. Under a strong magnetic field, the free electrons in the semiconductor do cyclotron resonance to radiate THz waves. The blackbody radiation source is a commonly used broadband THz source, and to obtain narrow-band THz radiation, a series of THz low-pass and band-pass filters should be placed in a metal tube coupling device. In the test system, to avoid the interference of stray light, it is necessary to place the visible light and infrared filter in the coupled light path and apply the blackbody material on the sample cavity wall to absorb the stray THz radiation.

B. **QD THz single-photon detector**

(a) **Structure and requirements of the QD THz single-photon detector**

QDs are semiconductor nanoparticles, only a few nanometers in size, bound in three dimensions. They exhibit unique optical and electrical properties due to quantum effects, distinguishing them from bulk materials. In 2000, Komiyama et al. at the University of Tokyo developed a highly sensitive THz single-photon detector by combining QDs with single-electron transistors (SETs). An SET, a nanoelectronic device, utilizes the Coulomb blocking and single-electron tunneling effects. The core of the SET, as illustrated in Fig. 3.44a, is a QD comprising two weakly coupled tunnel junctions and macroscopic electrodes. The SET is extremely sensitive to changes in

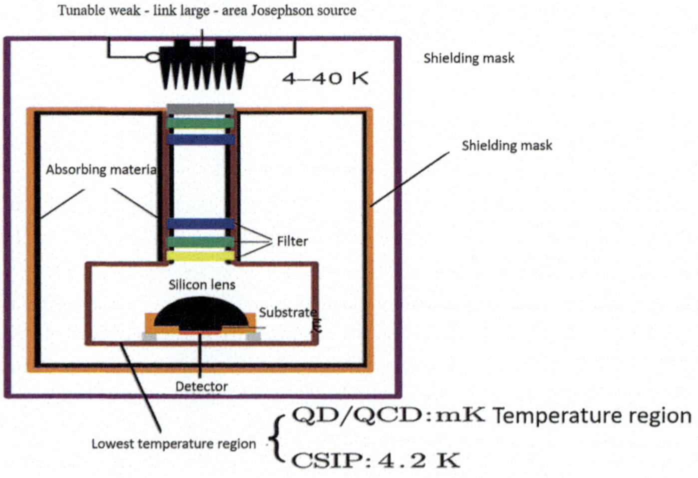


**Fig. 3.43** Schematic diagram of the low-temperature testing system for THz single-photon detector

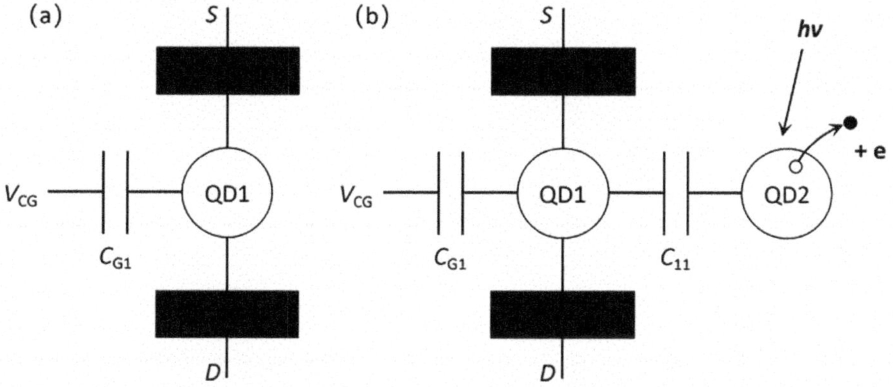


**Fig. 3.44** **a** SET diagram; **b** QDD diagram

the surrounding electric field. When a QD absorbs THz photons, electrons transition, polarizing the QD. The adjacent SET detects the resulting electric field changes, generating an induced current to enable photon detection. As depicted in Fig. 3.44b, QD 2 absorbs photons, while the SET, with QD 1 as its core, detects the electric field changes caused by QD 2 absorbing photons.

The design of QDDs relies on specific conditions to ensure their functionality. First, the thermal fluctuation must be much less than the charge energy of a single electron in a QD, maintaining a definite number of electrons in each QD. Second, the source and drain must be small, and the potential energy they provide must be much less than the charge energy to prevent excess electron injection into QD 1. Third, the energy of a single photon must exceed the energy level difference of QD 2, allowing QD 2 to transition after absorbing photons.

In the original design, the QDD adopted a single QD structure. Under a magnetic field bias, a Landau level forms, dividing the QD into two conductive regions: the core and the outer ring. This structure is equivalent to the coupling of a QD and a single-electron transistor (SET), forming the QDSPD. In 2001, Komiyama's team improved this design by introducing a Double QDSPD, which incorporated two QDs for enhanced performance.

(b) **Detection mechanism of the QD single-photon detector**

A magnetic field is applied perpendicular to the QD plane, separating the energy states of electrons into discrete Landau levels. Under appropriate magnetic field bias, two Landau levels are formed: the lowest level (LL 0), filled with electrons, and the first excited state level (LL 1), occupied by a smaller number of electrons. This configuration creates two conductive regions within the QD: the outer ring (corresponding to LL 0) and the kernel (corresponding to LL 1), as depicted in Fig. 3.45a. These regions are spatially separated and can be considered as two weakly coupled QDs. The kernel corresponds to QD 2 in Fig. 3.45b, while the outer ring corresponds to QD 1. The outer ring, along with its tunnel junction and electrode, forms a single-electron transistor (SET). By adjusting the gate voltage, the potential

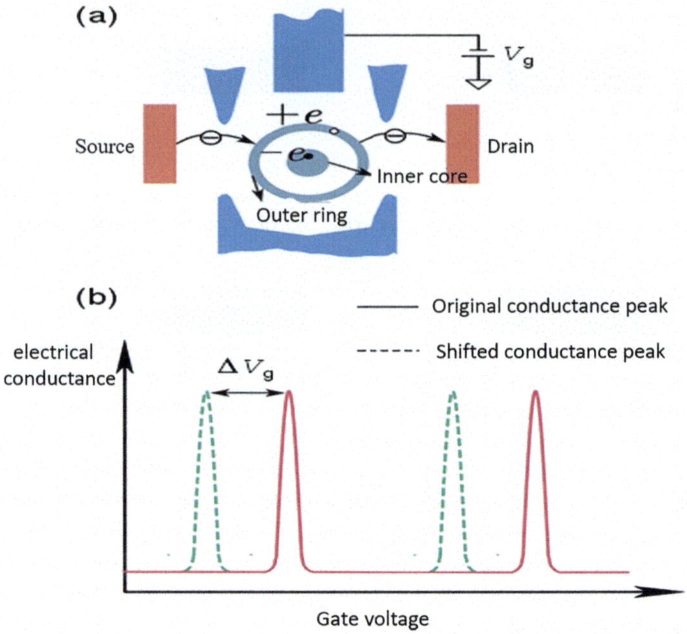


**Fig. 3.45** **a** The QD from the outer ring and the core under the strong magnetic field bias; **b** the shift of the conductance resonance peak

($\mu_0$) of the outer ring can be regulated. When $\mu_0$ equals the Fermi level (EF), a current flows from the source to the drain, causing conductance resonance peaks during gate voltage scans at $\mu_0 = EF$.

When THz photons are absorbed by the QDs, a hole forms in the low-level LL 0, and an electron occupies the high-level LL 1. This results in the core (LL 1) becoming negatively charged and the outer ring (LL 0) positively charged, causing electric polarization in the core. This polarization shifts the potential of the outer ring, leading to a conductance resonance peak shift in the direction of negative gate pressure. The offset ($\Delta V_g$) is proportional to the outer ring's potential change, enabling THz photon detection by monitoring conductance variations between the source and drain.

However, the single-QDD requires a strong magnetic field, limiting its practical application due to magnetic interference. The introduction of the double QD detector resolves this issue. In the double QD detector, shown in Fig. 3.46a, a negative bias is applied to the gate. This repels electrons in the lower 2D electron gas (2DEG), forming a depletion zone where the center is isolated to create a QD. Increasing the negative bias results in smaller QDs, enhancing performance and eliminating the need for a strong magnetic field.

In the double QDD, QD 2 absorbs incoming THz photons, as shown in Fig. 3.46b. This absorption excites electrons in QD 2, which then escape to the electron reservoir. A barrier between QD 2 and the reservoir ensures that the excited electrons cannot

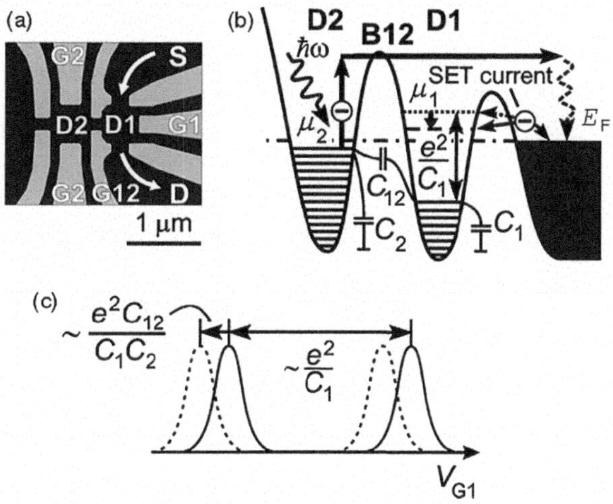


**Fig. 3.46** **a** schematic diagram of the dual QDD; **b** schematic diagram of photon absorption and electron transition; **c** resonance peak shift of the conductance caused by a single optical excitation event

return, facilitating efficient ionization. The single-electron transistor (SET) based on QD 1 detects the signal by monitoring changes in QD 2's ionization state. The principle of operation for the double QDD is similar to that of the single QDD. As depicted in Fig. 3.46c, the absorption of THz photons results in a shift in the conductance resonance peak. This shift corresponds to the polarization and ionization changes caused by photon absorption, enabling accurate detection of THz photons.

(c) **QD single photon detector readout circuit and single photon detection analysis**

QDSPDs require extremely low operating temperatures below 1 K. Traditionally, their readout circuits use phase-in amplification technology, which has a response time in the millisecond range due to the integration time of the lock-in amplifier. This slow response makes it challenging to distinguish single-photon absorption events in the time domain. To address this limitation, Ikushima et al. improved the readout circuit design. In their approach, the detector's current signal passes through a 40 kΩ resistor, converting it into a voltage signal. This voltage is then amplified, low-pass filtered, and recorded by an oscilloscope, as shown in Fig. 3.47a.

The new readout circuit significantly accelerates response time, reducing it to 20 μs. The conductance pulse signals from single-photon absorption events can now be observed in the time domain, as shown in Fig. 3.47b. Figure 3.47c illustrates the changes in the response signal before and after THz irradiation: the conductance cutoff corresponds to photon absorption, and its restoration indicates photocarrier recombination. By analyzing these conductance changes over time, the number of THz photons absorbed can be determined, enabling single-photon counting.

QDSPDs exhibit remarkable sensitivity. Their Noise Equivalent Power (NEP) depends on the ratio of the current noise spectral density to responsivity. The detectors operate at very low temperatures, minimizing system thermal noise. With a

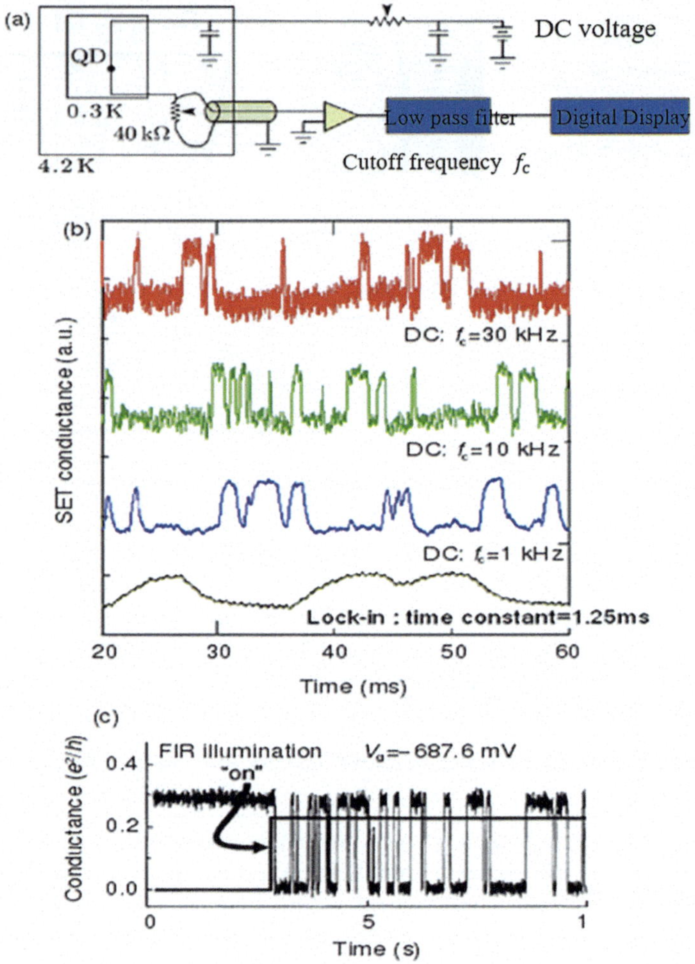


**Fig. 3.47** **a** QDD readout circuit; **b** test results of four different readout schemes; **c** the conductance-time trajectory of a single QDSPD

photoconductivity gain as high as $10^{12}$, QDDs demonstrate exceptional responsiveness. However, due to the high resistance of QDs and impedance mismatch with THz coupling structures, their quantum efficiency ($\eta_q$) is low. At 0.3 K, QDDs achieve a DCR of 0.1 beats per second, corresponding to an NEP of $10^{-20}$–$10^{-21}$ W/Hz$^{1/2}$, and a linear dynamic range exceeding 30 dB. At even lower temperatures (< 50 mK) and under magnetic field bias, the DCR drops to 0.001 beats per second, and the NEP improves to $10^{-22}$ W/Hz$^{1/2}$ with a quantum efficiency of approximately 1%. For dual QD detectors, the DCR reaches 0.01 beats per second, $\eta_q$ exceeds 0.1%, and the NEP is in the order of $10^{-21}$ W/Hz$^{1/2}$.

## C. Charge-sensitive infrared phototransistor (CSIP) detector

### (a) Operating principle of the CSIP detector

The QDSPD offers high sensitivity but is limited by the need for extremely low operating temperatures below 1 K, a complex fabrication process, and poor consistency, which restrict its broader application. To overcome these challenges, in 2005, the Komiyama research group introduced the charge-sensitive infrared phototransistor (CSIP) detector, based on a semiconductor dual quantum well structure. Quantum wells are potential wells for electrons or holes, formed by alternating layers of semiconductor materials with different band gaps, creating heterostructures that exhibit significant quantum confinement effects. The CSIP detector operates on a similar principle to the QDD, but with the advantage of being able to achieve highly sensitive detection at liquid helium temperatures. For the CSIP detector to detect THz single photons, it must meet two conditions: first, the change in charge should result solely from photon absorption, with thermal fluctuations having a minimal effect on the number of charges; and second, the absorption of THz photons must induce electron transitions, with the excited-state electrons having a long lifetime. Over time, the CSIP detector has evolved into two designs: the "vertical escape" structure and the "lateral escape" structure, both of which aim to enhance performance and detection efficiency.

The structure of the "vertical escape" CSIP detector is shown in Fig. 3.48a, consisting of two layers of quantum wells in a GaAs/AlGaAs heterojunction structure, with the band diagram shown in Fig. 3.48c. The difference in band gaps between GaAs and AlGaAs forms two quantum wells. Electrons are confined in a narrow potential well within the GaAs layer, known as a 2DEG. As illustrated in Fig. 3.48b, a comb gate is used to regulate the upper 2 DEG. By applying a negative bias voltage, the upper 2 DEG is separated into a series of conductive islets, which serve as the light-sensitive floating grid of the lower 2 DEG.

When incident THz photons are absorbed by a 2 DEG island in the upper layer, the electrons in the island transition and tunnel into the lower 2 DEG, increasing the conductance of the lower 2 DEG conductive channel. However, the electrons cross the conductive channel in a very short time, with a typical crossing time ($\tau_{\text{trans}}$) of 20 ps. Once the electron crosses the lower conductive channel, the photosensitive effect is

$$\Delta I_e = \frac{\alpha e}{\tau_{\text{trans}}}.$$

This is the geometric factor of the photosensitive zone. Generally, (1) the composite time ($\tau_{\text{life}}$) of the optical excitation charge in the CSIP detector is very long, usually higher than 1 h. The photoconductivity gain (GPC) of the CSIP can be expressed as

$$G_{\text{PC}} = \frac{\tau_{\text{life}}\, \Delta I_e}{e} = \frac{\alpha \tau_{\text{life}}}{\tau_{\text{trans}}}$$

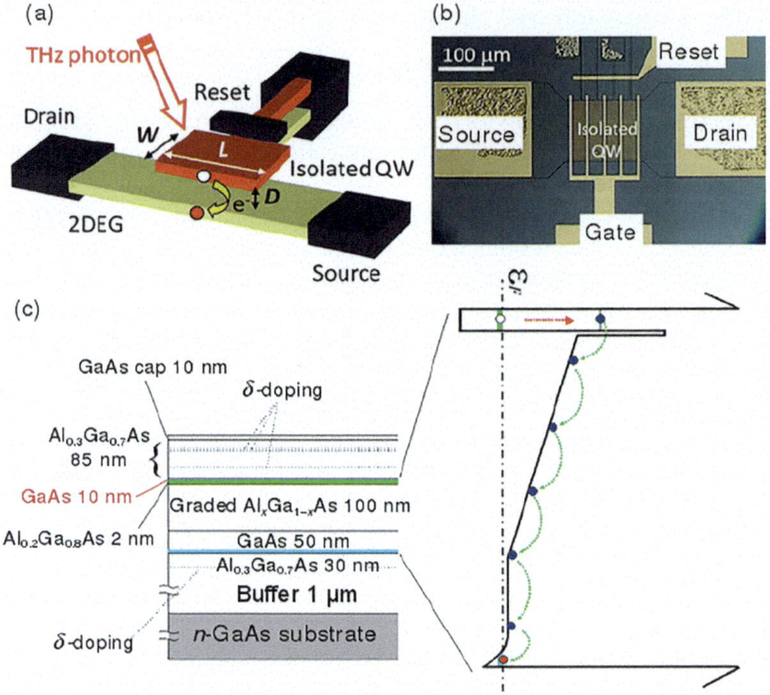


**Fig. 3.48** **a** Structural diagram of CSIP detector; **b** optical microscopy of CSIP detector; **c** GaAs/AlGaAs double quantum well heterojunction

Since τ life is far larger than $\tau_{trans}$, CSIP has an extremely high GPC, thus giving the CSIP detector the sensitivity to detect a single photon.

The "vertical escape" structure CSIP detector, while effective, is less sensitive in the long-wavelength segment. To address this limitation, Wang et al. proposed a "lateral escape" structure CSIP detector in 2012, as shown in Fig. 3.49. In this design, as depicted in Fig. 3.49a, b, a barrier is formed between the photosensitive floating gate and the electron reservoir by applying a negative bias voltage to the gate. Unlike the "vertical escape" structure, in which photoexcited electrons escape vertically, the "lateral escape" structure allows the photoexcited electrons to escape through the barrier into adjacent electron reservoirs.

The device utilizes a modulated-doped GaAs/AlGaAs heterojunction, as illustrated in Fig. 3.49c, d. The "lateral escape" structure CSIP shows improved performance over the "vertical escape" structure CSIP for wavelengths greater than 36 μm, offering better sensitivity in the long-wavelength segment.

Improving the quantum efficiency of the CSIP detector is a key focus of ongoing research. The efficiency of the device depends primarily on two factors: the efficiency with which the incident photon is absorbed by the light-sensitive floating gate ($\eta_1$) and the probability that the light-excited electrons escape from the floating gate into the lower quantum well ($\eta_2$). In the "vertical escape" structure CSIP, the initial

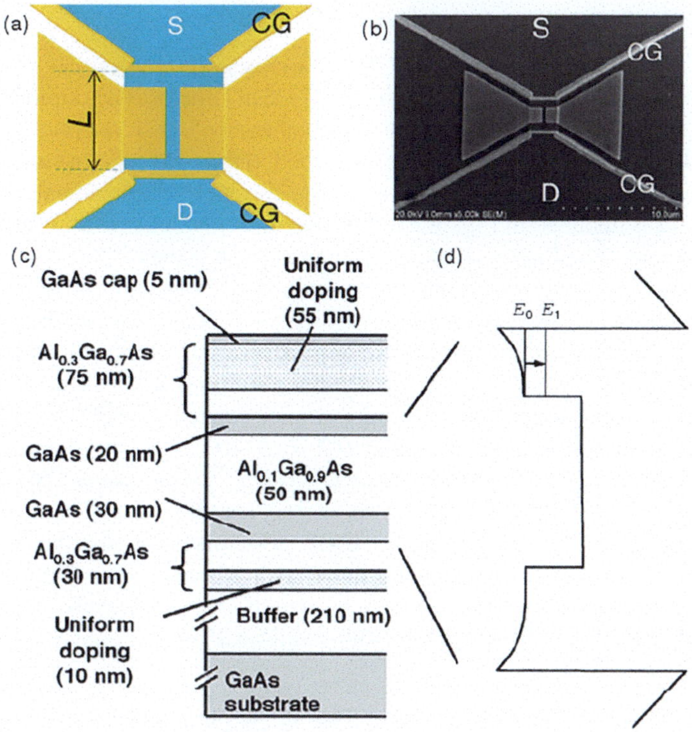


**Fig. 3.49** **a** "Lateral escape" CSIP structure; **b** CSIP actual device; **c** heterojunction material for "lateral escape" CSIP; **d** band diagram

optical coupling structure used was a square metal sheet array, which had relatively low coupling efficiency. Recently, this structure has been replaced with a Surface Plasmon Polariton (SPP) structure, which not only satisfies the polarization conversion requirements of the device but also provides field enhancement. By optimizing the SPP coupling structure, the quantum efficiency of the CSIP detector has been increased to 7%.

In contrast to the "vertical escape" structure, the "lateral escape" structure CSIP uses a butterfly antenna for the photocoupling structure, which offers better coupling efficiency. The factors influencing $\eta_2$ mainly include the quality of the heterojunction and the device structure. In the GaAs/AlGaAs heterojunction, impurity oxygen atoms introduced into the AlGaAs form the Al-O complex, which acts as a deep trap for electrons. In the "vertical escape" structure, photoexcited electrons must cross the barrier layer of AlGaAs, and the presence of deep traps hinders the electrons' transition to the lower 2 DEG, thus reducing $\eta_2$ and the sensitivity of the detector. However, in the "lateral escape" structure, the electron escape process occurs in the GaAs layer of the upper quantum well, which lacks Al atoms, so the escape process is not affected by the deep traps. This leads to improved electron escape efficiency and overall sensitivity.

**(b) The THz single-photon test of the CSIP detector**

The signal readout circuit of the CSIP detector closely resembles that of the QDD. Both "horizontal escape" and "vertical escape" CSIP structures use similar methods to read out and respond to signals. In the "vertical escape" CSIP, the photon response signal works as follows: as photons are absorbed, the positive charge on the photosensitive floating gate accumulates, causing the current response signal to increase. This behavior is depicted in Fig. 3.50a, b, where the conductance-time curve shows step signals corresponding to single-photon absorption events.

Due to the long lifetime of light carriers in the CSIP detector, the photosensitive floating gate remains positively charged, requiring a reset operation to prevent charge saturation. A short-period positive pulse applied via the "reset" electrode releases accumulated charge and restores the detector to a sensitive state. Figure 3.50c illustrates the current response signal before and after resetting. The "vertical escape" CSIP achieves high current responsiveness ($R = 10^5$–$10^8$ A/W), high sensitivity, and a noise-equivalent power (NEP) of $7 \times 10^{-20}$ W/Hz$^{1/2}$. Traditional CSIP detectors with two quantum well layers detect a narrow wavelength range (12–18 μm). Recent advancements include multi-frequency CSIP detectors with multilayer quantum wells. In 2015, Kim et al. developed a three-layer quantum well CSIP capable of highly sensitive detection of electromagnetic signals at a 5 μm wavelength. By 2017, Kim's team introduced a four-layer quantum well CSIP, further enhancing detection capabilities.

D. Superconducting quantum capacitor THz single-photon detector

The Quantum Capacitance Detector (QCD) is a THz single-photon detector based on a Single Cooper-pair Box (SCB). In 2009, Shaw et al. from the University of Southern California proposed the QCD concept, leveraging the sensitivity of SCB to alignment particle tunneling. Theoretical analysis indicated an upper NEP limit of $10^{-21}$ W/Hz$^{1/2}$.

**(a) SCB and quantum capacitor effects**

First, give a brief introduction to how SCB works. SCB is the basic unit of charged qubits (QBs) in the superconducting quantum computing circuit. Its structure is shown in Fig. 3.51a. The very small superconducting island is connected to the electron library through a superconducting tunnel junction, and the gate is used to control the tunneling of electrons from the electron library to the superconducting island. The Hamiltonian of the whole system is

$$\mathcal{H} = 4E_{\mathrm{C}}\left(n - n_{\mathrm{g}}\right)^2 - E_{\mathrm{J}}\cos\theta$$

where n is the number of excess Cooper pairs in the superconducting island, and $n_g$ represents the number of superconducting electron pairs under the gate voltage bias, which satisfies $n_g = C_g V_g/(2e)$ and is the phase difference of the Josephson junction. When the $E_J$ is ignored, the energy of the system takes on a parabolic form:

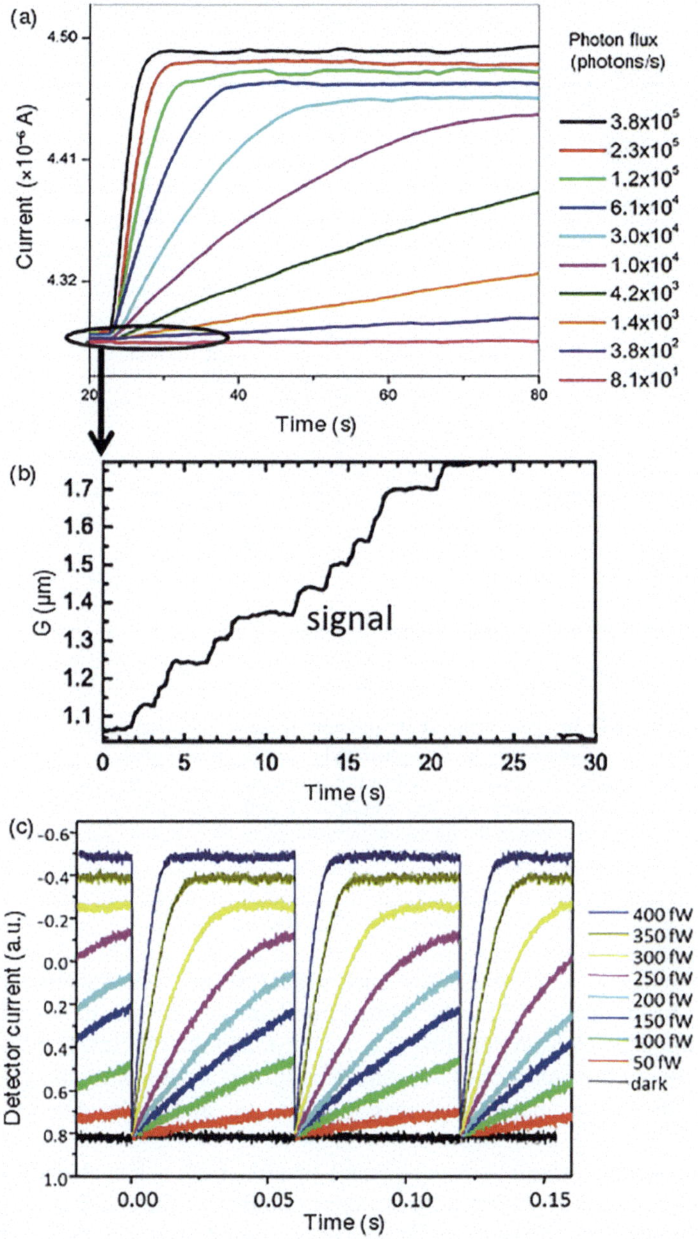


**Fig. 3.50** **a** CSIP current–time curve obtained under different irradiation conditions; **b** amplification diagram of the photoconductivity time trajectory at the lowest optical radiation power; and **c** photocurrent signal at different radiation power during reset operation

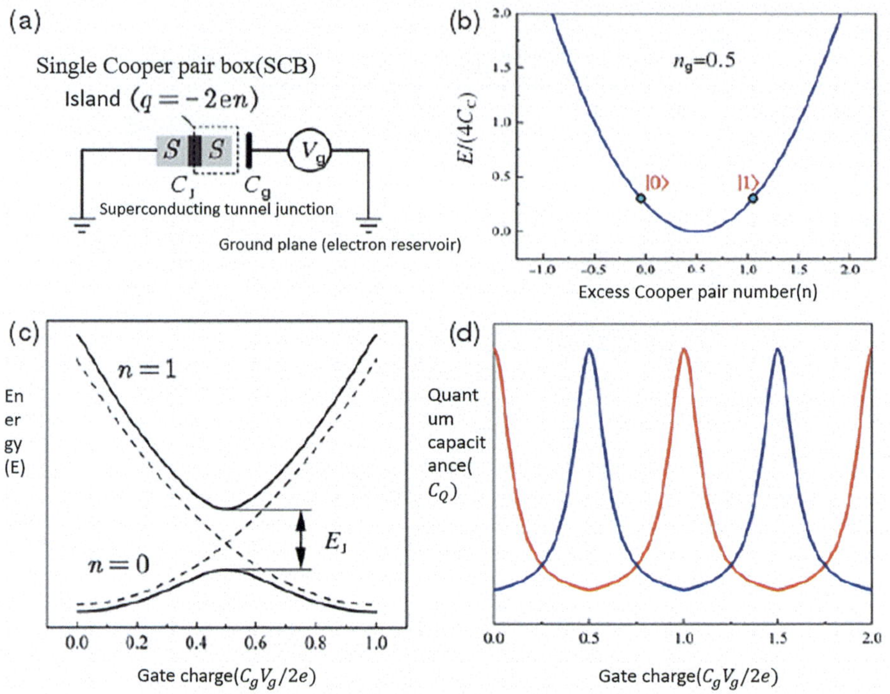


**Fig. 3.51** **a** Schematic diagram of the SCB structure; **b** the relationship between the normalized energy and the excess Cooper log in the superconducting island ($E_J = 0$); **c** Schematic diagram of the energy level under the approximation of the two-level system (solid line). The dashed line shows the schematic diagram of energy levels in (b) at $E_J = 0$, n = 0 and n = 1, both crossing at $n_g = 0.5$; **d** the relationship between the quantum capacitance of SCB and the gate voltage

$$E = 4E_C(n - n_g)^2$$

As shown in Fig. 3.51b, the lowest energy states 0 and 1 are degenerate when $n_g = 0.5$. As shown in Fig. 3.51c, the dashed line indicates the energy levels at 0 and 1 states without Josephson coupling (i.e. $E_J = 0$), and the two levels degenerate at $n_g = 0.5$. Due to the presence of Josephson coupling, as shown by the solid line in Fig. 3.51c, the two originally equal energy levels split, and the level spacing is $E_J$ at $n_g = 0.5$.

Superconducting islands are very small in size, with a low density of states, and have discrete energy levels. Due to the limitation of the Pauli incompatibility principle, the packing of the Fermi level can be caused from low to high level. For the external circuit, this effect can be equivalent to a capacitor, which is the quantum capacitance effect. The quantum capacitance is proportional to the second derivative of the energy level and the gate voltage. At the energy level degenerate point, the quantum capacitor is at the maximum.

Figure 3.51d shows the quantum capacitance curve of 0 and 1 quasiparticles tunneling into the superconducting island, which shows that the quantum capacitance changes periodically with the voltage bias. If the control gate voltage biases the superconducting island at ng = 0.5, the quantum capacitance changes significantly when a quasiparticle is tunneling into the superconducting island through the superconducting tunnel junction. It shows that the tunneling of the SCB is very sensitive, and the quasiparticle tunneling probability depends on the density of the quasiparticles in the superconductor, so the SCB can be used for the measurement of the quasiparticle density.

(b) **Working principle of QCD**

When THz photon energy exceeds the superconducting energy gap, the absorbed photons disrupt Cooper pairs, creating quasiparticles whose density correlates with the number of absorbed photons. For instance, when a 1.5 THz wave interacts with superconducting aluminum, approximately 20 quasiparticles are generated per photon. These quasiparticles tunnel into a superconducting island, altering its quantum capacitance, a principle utilized in Quantum Capacitance Detectors (QCDs) for highly sensitive THz detection.

QCDs employ Single Cooper-pair Boxes (SCBs) as their core components. Superconducting absorption gates or planar dipole antennas act as THz photon absorbers, while RF resonators coupled to SCBs measure quantum capacitance changes. The absorption gate, a lattice structure of aluminum nanowires, breaks Cooper pairs upon photon absorption, producing quasiparticles. These quasiparticles tunnel through the superconducting island, shifting the quantum capacitance and modifying the resonant frequency of the RF resonator. Each tunneling event generates a corresponding phase-hopping signal in the readout circuit. Since quasiparticles have a long lifetime compared to their tunneling time, multiple tunneling events can be observed following photon absorption. The tunneling probability depends on the quasiparticle density, enabling THz radiation detection through phase-hopping signal measurements.

For direct THz detection, the gate bias is scanned across multiple quantum capacitance cycles rather than fixed at the degenerate point (e.g., $n_g = 0.5$). The scanning signal frequency is lower than the tunneling rate, typically tens to hundreds of hertz, to mitigate low-frequency 1/*f* noise. Upon THz radiation absorption, quasiparticles in the absorption gate tunnel into the superconducting island as the bias voltage is scanned, producing a periodic signal corresponding to the tunneling events. The average value of the peak quantum capacitance is:

$$\overline{C_Q} = \left(\frac{C_g}{C_\Sigma}\right)^2 \left(\frac{2e^2}{E_J}\right) \frac{\Gamma_{out}}{\Gamma_{in} + \Gamma_{out}^1}$$

The rate of quasiparticle tunneling from the superconducting island, $\Gamma_{out}$, remains constant and does not depend on the number of quasiparticles in the absorption gate. Conversely, the rate of quasiparticles entering the superconducting island, $\Gamma_{in}$, is

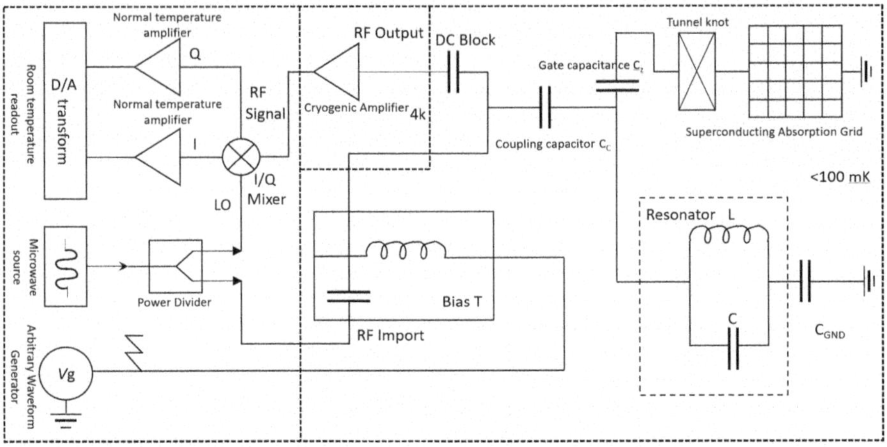


**Fig. 3.52** Schematic diagram of the structure and readout circuit of the QCD

proportional to the number of quasiparticles in the absorption gate ($N_{qp}$). Consequently, the height of the quantum capacitance curve is determined by $\Gamma_{\text{in}}$, which reflects the number of quasiparticles in the absorption gate. Since the number of quasiparticles depends on the number of absorbed THz photons, the power of the THz radiation can be measured by assessing the amplitude of the quantum capacitance.

(c) **Read-out circuit and SPD of the QCD**

The readout circuit of the Quantum Capacitance Detector (QCD), as shown in Fig. 3.52, involves biasing the QCD resonator using a microwave signal fed through the feeder. Quasiparticle tunneling causes a shift in the resonant frequency, generating a phase shift signal. This output signal is amplified by a low-temperature amplifier and converted to a voltage signal proportional to the phase shift using an IQ mixer. The voltage signal is then digitized using an analog-to-digital converter, and the data is processed by a computer for analysis.

When a single THz photon is absorbed, multiple quasiparticles are produced, resulting in multiple phase-hopping signals due to quasiparticle tunneling before their recombination. Ideally, phase-hopping signals only occur upon photon absorption. However, practical tests reveal a background tunneling rate at very low incident power, attributed to noise from stray radiation. To achieve SPD, interference from background tunneling events must be filtered out.

Guillaume et al. at Caltech developed a suppression method where the gate voltage scanning frequency exceeds the quasiparticle tunneling rate, eliminating tunneling in the absence of quasiparticles. By ensuring the scanning frequency is higher than the background tunneling rate but lower than the photon-induced tunneling rate, background events are suppressed, and photon absorption changes the quantum capacitance.

Echternach et al. demonstrated this approach by testing QCD responses under various irradiation conditions. At a 22 kHz gate scanning frequency and a sampling interval of 0.5 μs, data were grouped into 30 μs intervals, and the variance of the quantum capacitance signal was analyzed. Photon absorption events manifested as peaks in the variance–time curve, and time intervals between events followed a Poisson distribution. A histogram of photon arrival intervals (Fig. 3.53c) and the experimental probability distribution of photon absorption events (Fig. 3.53d) aligned with the Poisson model. These results confirm that QCD can detect single photons at 1.5 THz.

E. **Other THz single-photon detectors**

(a) **Nanometer thermal bolometer**

The nanometer is a highly sensitive THz detector with a theoretical value of NEP up to 10 when the operating temperature is less than 0.1 $K^{-20}W/Hz^{1/2}$, showing

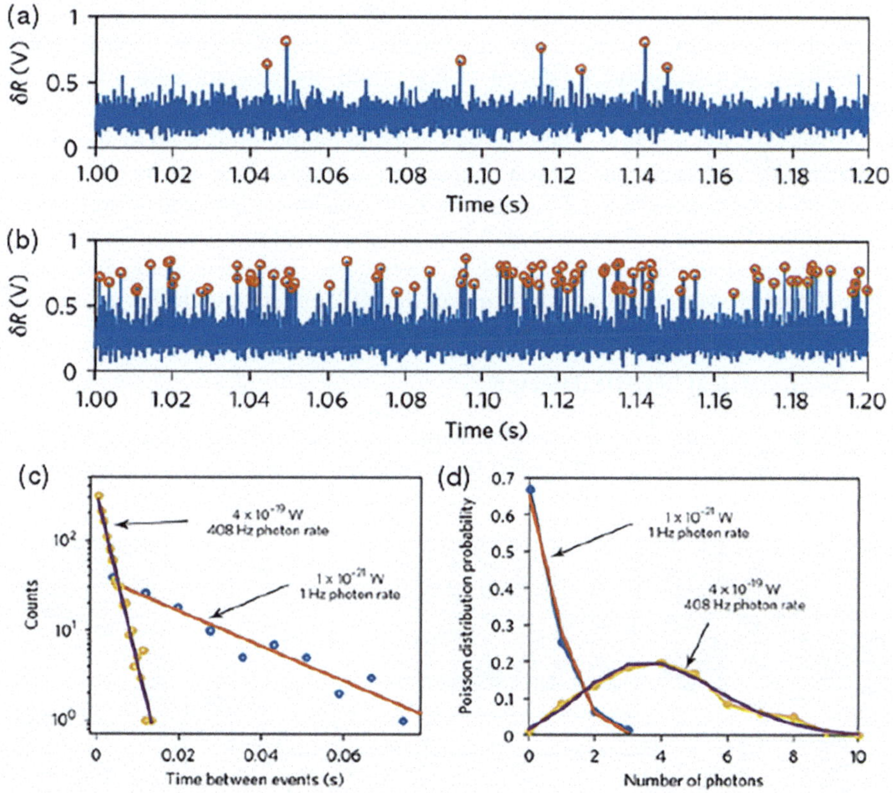


**Fig. 3.53** Radiation power is $1.13 \times 10^{-21}$ W (**a**) and $4.1 \times 10^{-19}$ W (**b**) the trajectory diagram of the variance of the quantum capacitor response signal of QCD over time, the red circle represents the peak; **c** the time interval between photon arrival events at two different radiation powers, the solid line is merged according to the Poisson distribution to obtain the photon arrival rate; **d** the measurement probability (circle) with an integer number of photons in the 36 ms interval, and the solid line is the calculated value using the Poisson distribution model

the prospect of application to THz single photon detection. In the thermal electron bolometer, the coupling between the electrons and the phonon system is very weak, and the absorbed photons increase the temperature of the electrons in the absorption unit, leading to an increase in electrical resistance. The smaller the heat capacity of the absorption unit of the nanometer meter, the better the thermal isolation from the surrounding environment, and the higher the sensitivity of the detector.

Rutgers University Wei et al. designed and prepared a thermal electron nanometer with Ti nanobridge. By improving the device design, they effectively suppressed the diffusion of electrons and the emission of photons, thus reducing the heat exchange of thermal electrons and the surrounding environment. Meanwhile, due to the small size of the nanobridge and its extremely low heat capacity, the detector obtains extremely high sensitivity. Based on the test results in the NIR band, the NEP is 910 at 1.9 THz at a temperature of 65 $mK^{-21}W/Hz^{1/2}$. Whether the detector has the THz single-photon counting capability still needs to be further experimental verification.

In 2019, a novel nanothermometer with a NEP of up to $210^{-20}W/Hz^{1/2}$, Has the potential for THz SPD. The device is small and its extremely low heat capacity gives the device high sensitivity. As shown in Fig. 3.54b, the detector with a length of 10 μm and a width of 200 nm formed 8 SNS junctions with the Al electrode, namely the $Al/Au_xPd_{1-x}/Al$ junction. The first SNS junction on the left is used to absorb the THz radiation and cause an electron temperature increase in the nanowire. Several other SNS are structured into LC resonators for the readout of the signal. The readout circuit of the detector is shown in Fig. 3.54a, using a Josephson parametric amplifier (JPA) for signal amplification. JPA has a lower noise coefficient compared to the conventional HEMT amplifier, which makes the NEP obtained with JPA an order of magnitude lower compared to the HEMT amplifier.

### (b) **Asymmetrical SQUID detector**

In 2018, a THz single-photon detector scheme based on the SQUID proximity effect was proposed by the Italian SPIN-CNR Institute Solinas et al. The device diagram and

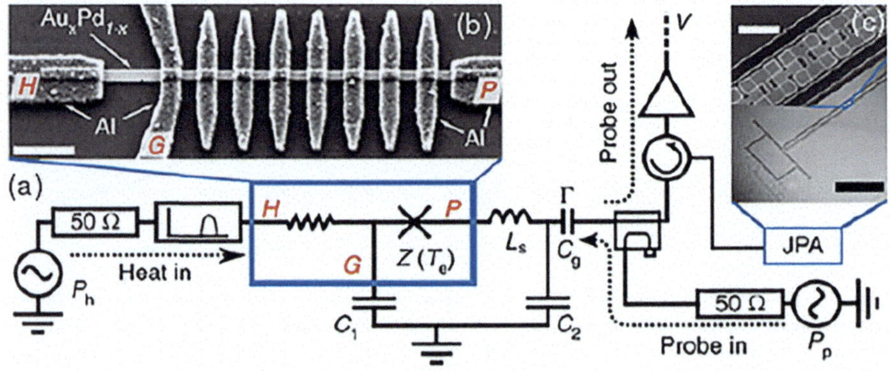


**Fig. 3.54** **a** Schematic diagram of nanometer readout circuit; **b** microscopy picture of nano bolometer; **c** microscope picture of Josephson Parametric amplifier (JPA)

working principle are shown in Fig. 3.55 to bias the two asymmetric SNS junctions in the unstable state by adjusting the magnetic flux in the superconducting ring. The radiated photons cause an increase in the electron temperature in the junction and eventually a sharp voltage signal from SQUID. SPD can be achieved by enhancing the voltage pulse signal induced by a single photon incident by selecting a suitable single photon parameter. The shape and amplitude of the voltage pulse are affected by the photon frequency, the junction asymmetry, and the operating temperature. Only the theoretical analysis and parametric simulation of the device performance are presented in the literature, and no relevant experimental work has been reported.

F. **Application of THz single-photon detectors**

(a) **Astronomical observation**

Since the Big Bang, the submillimeter or THz region of the electromagnetic spectrum accounts for 98% of all emitted photons, and the THz spectrum can reveal the laws of galaxy evolution and enrichment of heavy elements throughout the universe. This radiation is strongly absorbed in the Earth's atmosphere, so space-based THz telescopes are crucial for exploring the development of the universe, prompting astronomers around the world to develop space telescopes for imaging and spectral analysis of the THz frequency band. The low quantum efficiency of QDD and

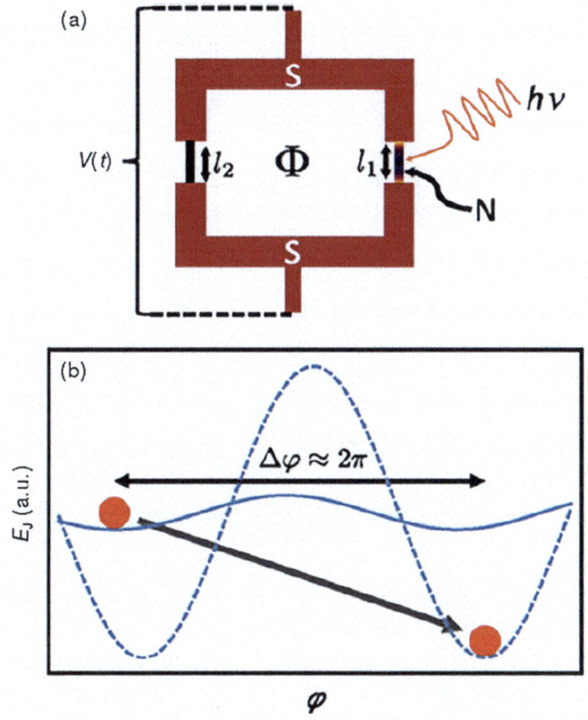


**Fig. 3.55** **a** Schematic diagram of an asymmetric SQUID radiation detector; **b** change of Josephson coupling energy before and after photon absorption, the solid and dashed lines are Josephson coupling energy before and after photon absorption respectively

CSIP detectors limits their application in astronomical observations. QCD has high quantum efficiency and relatively simple array readout techniques, with potential applications in future astronomical observations.

The sensitivity of the detector is a key factor limiting the performance of the THz spectrometer. Take the waveguide far-infrared spectrometer ($W_a$-FIRS) commonly used in astronomical observation as an example. This spectrometer is based on the principle of diffraction grating separation, and the sensitivity of the detector determines the spectral resolution of the spectrometer. The development of THz single-photon detectors will facilitate the improved resolution of grating spectrometers to meet the demands of astronomical observations. In addition to spectroscopic applications, THz single-photon detectors act as a high-sensitivity detector with promising promise in astronomical imaging applications. NASA's Origin Space Telescope imaging system requires the probe to have a NEP of $210^{-19}$W/Hz$^{1/2}$.

(b) **THz microscopy technology**

In conventional microscopy imaging systems, an external light source is usually required to image the transmission, reflection, or scattering of the incident light. Because the temperature will constantly radiate electromagnetic waves, radiation photons carry the information of the object, these photons are captured through the microscope and can be used for the study of localized phenomena inside the object, this is the passive microscopic technology. Because the thermal radiation of objects is often very weak, passive microscopy requires highly sensitive detectors. In 2006, Ikushima et al. of JST, Japan, built a THz photon counting microscope using QDD to study the THz radiation and non-equilibrium electron distribution in semiconductor quantum Hall devices. Because of the far-field imaging method, the spatial resolution of the microscope is about 50 μm. Then the spatial resolution of this technique can be further improved.

To break the diffraction limit, in 2018, Weng Qianchun et al., Shanghai Institute of Technical Physics, Chinese Academy of Sciences, developed THz scanning noise microscopy technology (SNoiM) based on CSIP detectors. The granular noise generated by the electron fluctuations of the non-equilibrium state radiates the near-field THz wave, and the near-field THz signal is detected through the CSIP detector to observe the characteristics of the non-equilibrium electron. SNoiM is a technique for passive near-field imaging with a spatial resolution of tens of nanometers. SNoiM technology can be used not only for the study of non-equilibrium electrons in semiconductor electronic devices but also for the study of non-equilibrium electronic systems in non-metallic or semi-metallic materials in the future.

(c) **Quantum communication and quantum radar**

Quantum communication technologies, particularly those based on quantum entanglement, have advanced rapidly, with QKD technology forming the backbone of secure communication. In QKD systems, the transmitted quantum state signal is extremely weak, and single-photon detectors are used to measure the quantum state at the receiving end. The performance of these detectors directly impacts the confidentiality, stability, transmission distance, and communication rate of the

QKD system. Currently, quantum communication typically operates in the 1550 nm wavelength range. However, with advancements in THz (THz) SPD, THz-frequency quantum communication technology is emerging, offering a new avenue for secure communication in challenging environments, such as those with soot or smoke.

In parallel, THz radar systems are becoming a key area of radar technology development. THz radar offers higher spatial resolution compared to microwave radar, enabling the detection of smaller targets. Unlike lidar, THz waves can penetrate smoke and dust, making them ideal for searching for space targets. THz single-photon detectors can significantly enhance the accuracy of THz radar target detection, potentially advancing the development of THz quantum radar technology.

(d) **Single-photon imaging**

The development of SPD technology has promoted the improvement of the action distance of laser range measuring and radar imaging. When the imaging distance is tens of kilometers or longer, only a few photons can return to the detection system, which requires the high sensitivity of the detector. The accurate measurement of the time of flight of photons is of great significance in architecture, defense science, and earth science research. Currently, single photon imaging has achieved 45 km in the 1550 nm band. At present, single photon imaging in the THz band has not been reported. With the development and maturity of related technologies such as THz single photon detection, THz single photon imaging technology will also become an important development direction of THz imaging technology in the future.

G. **Development trend of THz single-photon detectors**

(a) **Large-scale detector array**

The current number of THz detector arrays is small compared to optical and near-infrared detector arrays. Large-scale THz detector arrays help to improve the field of view and observation speed of imaging systems such as telescopes. To meet the increasing demand for astronomical observation, the number of pixels in the detector array is increasing. Currently, THz direct detector arrays containing 105 pixels are in development.

(b) **The detector readout circuit**

The implementation of a large-scale detector array will lead to the complexity of readout circuits and an increase in power consumption. High-performance low-noise cryogenic amplifiers are a problem to be solved in the future. Recently, JPA has been widely used in quantum computing circuits, and the noise temperature of JPA is lower compared to traditional low-temperature amplification techniques. JPA's working bandwidth is still narrow, and for a large-scale THz detector array, developing a broadband adjustable JPA is also a technical challenge.

(c) **Dynamic range**

For space observations, the background noise is low, which has more stringent requirements on the detector's dynamic range, usually several orders of magnitude higher than ground-based observations. To meet the needs of space observations, the dynamic range of the detector array needs to be increased by at least five magnitudes in the future.

In terms of the sensitivity of the detection, QDD has the highest sensitivity, with the NEP of a single QDD reaching up to $10^{-22}$W/Hz$^{1/2}$ and a double QDD of up to $10^{-21}$W/Hz$^{1/2}$. The QDD has a low quantum efficiency (< 1%), which seriously limits its application range. The CSIP detectors have an NEP of up to $10^{-21}$W/Hz$^{1/2}$. Compared to QDD, the CSIP detector has a wider dynamic range (> 106), a higher operating temperature (> 4.2 K), and a higher quantum efficiency (1–7%). Moreover, the CSIP detector structure is simple to realize large-scale detector arrays and on-chip readouts, so the application is more promising. Superconducting QCD has now achieved SPD at 1.5 THz, with a quantum efficiency of about 90%. The main advantage of QCD is the simple readout circuit and easy preparation of the detector array. The main factor limiting its development is the difficulty in preparing SCB. Compared with the theoretical limit, the performance of the THz single-photon detector still has some room for improvement. Properly reducing the volume of the detector photosensitive unit, improving the quality of the film, and using a low-noise amplifier such as JPA can all help to improve the detector sensitivity. For THz single-photon detectors, the device quantum efficiency can be improved by optimizing the coupling structure design and improving the device quality. For a SPD, the detection photon signal is measured in a mode of photon counting, and the dynamic range of the detector depends on the counting rate of the detector. We can effectively improve the dynamic range of the single-photon detector by optimizing the dead time and the integration time of the readout circuit.

## 3.3 Performance Evaluation and Optimization of Single-Photon Detectors

1. **Performance evaluation of the single-photon detector**

(1) **Performance indicators**

A. **Detection efficiency**

Detection efficiency is a key performance indicator for single-photon detectors. A higher detection efficiency means the detector is better at capturing incident photons and utilizing their energy effectively. Several factors influence detection efficiency, including the optical absorption coefficient of the detector, the transmission efficiency of photogenerated electrons, and the optical interface reflectivity.

B. **Dark count rate**

The DCR refers to the counting rate of a detector when no light signal is present. It results from internal noise, such as thermal and electronic noise within the detector.

A lower DCR indicates better noise performance and higher sensitivity to optical signals. In performance evaluation, minimizing the DCR is crucial to improving the SNR of the detector, ensuring more accurate detection of optical signals.

C. **Temporal resolution**

Temporal resolution refers to the minimum time interval at which a detector can distinguish two consecutive photon events. A higher temporal resolution means the detector can respond more quickly to optical signals and better capture the timing details of photon events. Factors such as circuit noise, the photoelectric conversion process, and signal processing technology influence temporal resolution.

D. **Stability and reliability**

Stability and reliability are crucial indicators for assessing the long-term performance of a detector. Stability refers to the consistency of the detector's performance over time, including its detection efficiency, DCR, and other key metrics. Reliability, on the other hand, measures the detector's ability to function stably under specified operating conditions and maintain good performance after extended use or a set number of triggers.

E. **Dynamic range and linearity**

Dynamic range refers to the range of signal intensities that a detector can process, while linearity refers to the direct proportionality between the output signal of the detector and the input signal. A larger dynamic range allows the detector to handle optical signals of varying intensities, while better linearity ensures that the output signal closely matches the input signal, providing a more accurate reflection of the optical signal's intensity.

(2) **Performance evaluation methods**

When evaluating the performance of a single photon detector:

A. **Laboratory testing**

To properly test a single-photon detector, it is essential to first prepare the test environment by ensuring it is clean, stable, and free from dust to prevent external interference. Additionally, room temperature and humidity should be adjusted to meet the required standards for accurate testing. Before operating the detector, its surface should be cleaned using the appropriate method to remove any contaminants or impurities that might affect its performance. Once the detector is cleaned, it should be connected to the necessary test equipment, such as a power mete, and optical waveform instrument, ensuring stable and accurate connections. To test detection efficiency, a standard light source should be used, directing its output to the detector. The output signal should then be recorded and analyzed to calculate the detection efficiency. During this process, an optical power meter is essential for testing the light intensity and maintaining the consistency of the input light intensity to the detector.

B. **Practical application scenario test**

To assess the actual performance of a single-photon detector in a specific application, such as QKD, it is essential to consider various factors. In QKD, one of the key requirements is ensuring the real-time randomness of the photons, which can be achieved by evaluating the detector's dynamic range for the signal from the laser and its ability to accurately identify the laser frequency. Additionally, the detector's source power, wavelength, structure, and response time must be carefully considered to ensure optimal performance. These factors should be verified through experimental methods to validate the accuracy and reliability of the detector in the given application scenario.

C. **Data analysis and comparison**

In addition to detection efficiency, several other factors must be considered when evaluating the performance of single-photon detectors, including the DCR, temporal resolution, and spectral response. The DCR refers to the spurious triggers caused by spontaneous circuit noise in the detector when no light source is present. This can be minimized by adjusting the detector's temperature and optimizing the design of the light detection region. Temporal resolution is critical for both the precision of the detector and its suitability for high-precision measurements, and it can be assessed through techniques like time-of-flight measurements. Spectral response, which indicates how responsive the detector is to photons at different wavelengths, is another key performance indicator. To comprehensively evaluate a single-photon detector's performance, data from experimental testing and practical application scenarios should be compared and analyzed, taking into account all of these factors.

2. **Optimization strategy for the single-photon detector**

(1) **Improve the detection efficiency**

A. **Optimized material selection**

To improve the efficiency of photon conversion into detectable signals, materials with high quantum efficiency should be selected. These materials can more effectively convert absorbed photons into electrical signals, thereby enhancing detection efficiency. For instance, using materials with high absorption coefficients and low attenuation rates can increase the absorption and conversion efficiency of photons. Additionally, selecting materials with high exciton binding energy helps reduce exciton recombination, further improving the probing efficiency and ensuring better performance of the detector.

B. **Improving device structure**

The goal is to enhance photon-detector interactions by designing more effective optical microcavity structures or developing multi-channel single-photon detectors. Optimizing the microcavity structure increases the photon cycle within the detector, boosting detection efficiency. Additionally, creating multi-channel detectors allows for improved integration and enables parallel processing of multiple photon signals, further enhancing detection efficiency.

C. **Using high-sensitivity detector**

PMTs can achieve detection efficiencies exceeding 95%, making them highly effective for detecting low-intensity optical signals. However, their high sensitivity can introduce noise, which may affect the accuracy and stability of detection. To maintain optimal performance, it's important to choose appropriate high-sensitivity detectors and implement noise reduction measures. This balance ensures both high detection efficiency and minimal noise interference.

(2) **Reducing dark count rate**

A. **Reduce the operating temperature of the detector**

Dark counts are false signals produced by a single-photon detector in the absence of light, primarily due to thermal noise within the detector. Reducing the detector's operating temperature effectively lowers thermal noise, thereby decreasing the DCR. This is because semiconductor materials in the detector can generate noise when they absorb heat during operation, which may be misinterpreted as photon signals. Using cooling methods like liquid nitrogen or mechanical cooling improves heat conduction efficiency, directly lowering the detector's temperature. Optimizing the detector's structure and materials can further reduce heat buildup. Experiments show that such measures can decrease the DCR by over an order of magnitude, significantly enhancing the detector's quantum efficiency and overall performance.

B. **Improving the shielding design**

In addition to thermal noise, external electromagnetic interference (EMI) can also cause dark counts in single-photon detectors. Improving the detector's shielding design is an effective way to reduce these disturbances and lower the DCR. EMI can originate from nearby electronic equipment, power lines, and high-frequency facilities, and these signals may be falsely detected as photon events. To mitigate this, a closed shielding shell made of high magnetic conductivity materials can be used to absorb or reflect electromagnetic waves. The detector's design can also include a filter circuit to suppress high-frequency interference, while optimizing the circuit layout can reduce internal signal disturbances. Effective shielding eliminates dark counts caused by EMI, enhancing the detector's stability and reliability.

(3) **Improving temporal resolution**

A. **Developing novel single-photon detector technology**

Single-photon detectors are essential in fields like optical communication, quantum information, and high-speed computing. The SPAD is a cutting-edge detector known for its high time accuracy, excellent SNR, and low power consumption. SPADs offer picosecond-level temporal precision, making them ideal for high-precision measurements. They operate at high voltage, where a photon entering the detector generates an electron–hole pair. This triggers an avalanche effect, creating many more electron–hole pairs. These pairs are accelerated by the electric field, eventually reaching the detector's bottom and producing an electrical signal. By measuring the arrival time of this signal, the photon's arrival time can be precisely determined, allowing for accurate timing measurements.

B. **Optimizing readout circuits and the signal processing algorithms**

For readout circuits, using high-speed, low-noise amplifiers can improve signal amplification, while high-speed, high-precision timers help measure the arrival time of signals with greater accuracy. Additionally, employing digital signal processing techniques can filter out noise and interference, enhancing signal recognition. Optimizing signal processing algorithms further improves temporal resolution. Techniques like time series analysis and template matching can refine the measurement of signal arrival times. Machine learning methods can also be applied to train models, predict, and eliminate noise, thereby enhancing the detector's overall performance.

(4) **Enhancing stability and reliability**

A. **Quality testing**

In the production of single-photon detectors, strict quality control is crucial to ensure their stability and reliability. This involves comprehensive performance testing of each component, such as PMTs, PDs, and circuit boards. Key parameters, including response time, detection efficiency, noise levels, and dynamic range, should be carefully evaluated. In addition to testing individual components, it is essential to conduct full system testing to assess the detector's performance under real-world operating conditions. These tests can be carried out in a simulated environment that mimics challenges like temperature fluctuations and changes in light intensity.

B. **Periodic calibration and maintenance**

These detectors require regular calibration and maintenance to ensure optimal performance, as the internal electronic components may degrade or age over time. Calibration involves adjusting parameters such as gain and threshold to restore the detector to its best working condition. Maintenance tasks may include replacing worn-out parts, cleaning optical windows, and checking circuit connections. To streamline calibration and maintenance, an automated detection system can be employed, allowing tests to be performed without human supervision and enabling automatic adjustments of the detector parameters. Additionally, maintaining a detailed log of each maintenance event—including the time, actions taken, and adjustments made—is crucial for tracking performance changes and predicting potential failures, ensuring the continued reliability of the detector.

## 3.4 Application of Single-Photon Detectors in Imaging

Single-photon detectors are devices capable of detecting single photon events, and they play an important role in imaging techniques.

1. **Operational mechanisms**

A single-photon detector is a sensitive photoelectric device that detects photons by creating an electron–hole pair upon photon interaction with the detector material. The process involves photon absorption, electron–hole pair generation, charge separation under an electric field, and the production of an electrical signal. Measuring this signal's intensity and timing allows determination of the photon's energy and arrival time.

These detectors are widely used in medical imaging techniques like SPECT and PET, as well as in high-energy physics, astronomy, and other research fields, owing to their high sensitivity and precision.

2. **Applications**

The applications of SPD in imaging technology are wide ranging, mainly including the following aspects:

(1) **Quantum imaging**

Quantum imaging is an advanced imaging technique leveraging quantum mechanics principles, including quantum entanglement and quantum interference. Single-photon detectors are essential in this technology, enabling the detection of individual photons at low light intensities and accurately measuring their arrival time and intensity. Quantum imaging has significant applications in confidential communication by enabling quantum encryption and secure transmission of optical signals. In biological imaging, it provides high-sensitivity imaging, producing clearer and more detailed visuals of biological tissues, advancing biological research. The integration of single-photon detectors in quantum imaging plays a crucial role in achieving high-sensitivity imaging and enabling impactful applications in secure communication and biomedical fields.

(2) **High-speed imaging**

Single-photon detectors are pivotal in high-speed imaging due to their exceptional temporal resolution, enabling precise imaging of rapidly moving objects. In scientific research, they support techniques like high-speed optical microscopy and spectral imaging, allowing real-time observation of fast processes, such as biological cell dynamics and chemical reactions. In industrial applications, they are used in high-speed photography for detecting and analyzing moving objects, enhancing production efficiency and quality control. Additionally, single-photon detectors facilitate high-speed communication by enabling accurate detection and processing of optical signals, improving communication system reliability and speed. Single-photon detectors are invaluable in high-speed imaging, contributing significantly to advancements in scientific research, industrial production, and communication technologies.

(3) **Low-light imaging**

Single-photon detectors excel in low-light environments where traditional imaging devices fail due to sensitivity limitations. By detecting individual photons and accumulating them, these detectors can create detailed images even in extremely low-light conditions. In astronomy, they are crucial for studying distant objects emitting minimal photons, enabling the detection of faint cosmic signals. In night vision monitoring, single-photon detectors can form clear static images under low light, surpassing traditional devices that struggle with such clarity. Beyond imaging, single-photon detectors find applications in high-speed communication and quantum computing. With ongoing technological advancements, their performance is expected to improve, broadening their use in low-light imaging and other fields.

(4) **Biological imaging**

Single-photon detectors are essential tools in biological imaging [26–30], offering high-resolution imaging of biological samples labeled with fluorescent molecules. This capability plays a vital role in biological and medical research. In fluorescence microscopy, single-photon detectors enable precise imaging of individual molecules, providing insights into biomolecular structures, functions, and mechanisms of action. They allow real-time monitoring of dynamic molecular changes in live cells, aiding studies in cell biology and pharmacology. These detectors are also used for imaging biological tissues, where their ability to detect fluorescence signals helps investigate tissue structures and functions. Applications extend to medical diagnostics, biological research, and beyond. Single-photon detectors are invaluable for studying biological samples, uncovering physiological and pathological processes, and advancing research in biology and medicine.

3. **Technical challenges**

Despite their significant potential in imaging technology, single-photon detectors face several technical challenges.

(1) **Improving the sensitivity of the detector**

The sensitivity of single-photon detectors, a crucial performance measure, is directly tied to their ability to detect individual photons. Researchers have implemented various strategies to enhance this sensitivity. One approach is the use of high-purity materials, such as Si or Ge, which reduces noise caused by impurities and improves the detector's quantum efficiency. Structural optimization is another key strategy; for instance, Backside Illumination (BSI) technology minimizes photon absorption and scattering losses by shortening the propagation path within the detector, thereby enhancing the sensitivity of PD. Additionally, advancements in readout circuits, such as time-integrated CCDs (TICCDs) and charge injection devices (CIDs) [31, 32], significantly improve the SNR during signal readout, further boosting the overall performance of single-photon detectors.

(2) **Reducing the noise**

Reducing noise is crucial to improving the performance of single-photon detectors, as background noise can significantly affect the accuracy of detection results. One effective strategy is to use low-noise amplifiers in the signal amplification process. For instance, co-base amplifiers or field-effect transistors (FETs) can minimize noise without compromising signal amplification. Additionally, specific optical filtering techniques, such as band-pass filters or interference filters, help eliminate unwanted light from external sources, reducing noise caused by ambient or stray light. Furthermore, improving the physical structure and materials of the detector can lower intrinsic noise levels. For example, using superconducting nanowire single-photon detectors (SNSPDs) or materials with reduced thermal noise can decrease the detector's intrinsic noise limit, enhancing its overall performance.

(3) **Extending the life of the detector**

The key to extending the lifespan of single-photon detectors in imaging technology lies in optimizing materials and structural design. Since these detectors are often exposed to high voltage and current, they experience significant wear, which impacts their longevity. To enhance their lifespan, selecting materials with high stability and good electrical properties is essential, as these can minimize performance degradation. For instance, using high-quality semiconductor materials can increase durability and reduce failure rates. Additionally, improving the detector's structural design—such as making it more compact—can alleviate physical and electrical stress, further extending its life. Using materials resistant to high pressure and temperature, like high-temperature plastics or ceramics, can also improve the detector's resistance to harsh operating conditions.

(4) **Realizing high-speed readout**

The readout speed of single-photon detectors is crucial for high-speed imaging applications. To achieve faster readout, researchers are focusing on developing advanced readout circuits and optical transmission technologies. Traditional circuits, such as CCD or CMOS technology, have limitations in high-speed applications. As a solution, new circuits like SPAD readout circuits are being developed for faster performance.

In addition, high-speed optical transmission technologies are being explored to enhance the speed of optical signal transmission, which directly impacts imaging speed. Fiber-optic technology is one such advancement, offering low signal attenuation and minimal noise interference. However, challenges remain, such as power consumption, noise in high-speed circuits, and signal attenuation in optical transmission. Researchers are working on optimizing circuit designs and transmission techniques to address these issues and improve both readout speed and imaging quality.

Despite the technical hurdles, the continued progress in technology suggests that these challenges will be overcome, and the potential for single-photon detectors in imaging technology is vast. As innovations advance, single-photon detectors are expected to play an increasingly important role in the future of imaging.

## References


1. Polyakov SV (2013) Photomultiplier tubes. Exp Methods Phys Sci 45:69–82. https://doi.org/10.1016/B978-0-12-387695-9.00003-2
2. You A, Be MAY, In I (1989) An experimental study of a metal-oxide-semiconductor photomultiplier. J Appl Phys 66(6):2678–2688. https://doi.org/10.1063/1.344237
3. Hynecek J (1992) CCM—a new low-noise charge carrier multiplier suitable for detection of charge in small pixel CCD image sensors. IEEE Trans Electron Devices 39(8):1972–1975. https://doi.org/10.1109/16.144694
4. Gu K, Liu X, Zhang Y, Zhao H, Liu W (2020) Application of vacuum photoelectric detection technology in super-resolution system. Opt Photonics J 10(06):141–148. https://doi.org/10.4236/opj.2020.106015
5. Foord R, Jones R, Oliver CJ, Pike ER (1969) The use of photomultiplier tubes for photon counting. Appl Opt 8(10):1975. https://doi.org/10.1364/ao.8.001975
6. InGaAsP-InP single photon avalanche diodes with ultra-high.pdf
7. Li J et al (2023) Design of a room-temperature, sine-wave gated, InGaAs/InP SPAD based photon counting system with dead-time mitigation. J Light Technol:1–7. https://doi.org/10.1109/JLT.2023.3332480
8. Yang B, Zhao Y, Chen J (2021) High sensitivity graphene-$Al_2O_3$ passivated InGaAs near-infrared photodetector. Nanotechnology 32:455503 (8pp)
9. Jeong H, Kim DW, Lee SH, Kim KH (2014) Measurement of single-photon detection efficiency of InGaAs/InP avalanche photodiodes in wavelength-division-multiplexed schemes. In: 2014 12th international conference optical internet, pp 1–2. https://doi.org/10.1109/COIN.2014.6950611
10. Donnelly JP et al (2006) Design considerations for 1.06-μm InGaAsP-InP Geiger-mode avalanche photodiodes. IEEE J Quantum Electron 42(8):797–809. https://doi.org/10.1109/JQE.2006.877300
11. Zhao Y, Chen J (2022) InGaAs/InP SAGCM avalanche photodiode with a heterojunction multiplication layer. Phys B Condens Matter 629:413637. https://doi.org/10.1016/j.physb.2021.413637
12. Franco DS et al (2005) High-performance InGaAs–InP APDs on GaAs. 17(4):873–874
13. Mohammad Nejad S, Aghaei F (2020) Noise characteristics improvement of submicron InP/InGaAs avalanche photodiode for laser detection system. Opt Commun 455(Sept 2019):124561. https://doi.org/10.1016/j.optcom.2019.124561
14. Wu D et al (2021) Ultrabroadband and high-detectivity photodetector based on $WS_2$/GE heterojunction through defect engineering and interface passivation. ACS Nano 15(6):10119–10129. https://doi.org/10.1021/acsnano.1c02007
15. Choi W, Choudhary N, Han GH, Park J, Akinwande D, Lee YH (2017) Recent development of two-dimensional transition metal dichalcogenides and their applications. Mater Today 20(3):116–130. https://doi.org/10.1016/j.mattod.2016.10.002
16. Ouyang T, Wang X, Liu S, Chen H, Deng S (2021) A complete two-dimensional avalanche photodiode based on $MoTe_2-WS_2-MoTe_2$ heterojunctions with ultralow dark current. Front Mater 8. https://doi.org/10.3389/fmats.2021.736180
17. Vu QA et al (2017) Tuning carrier tunneling in van der Waals heterostructures for ultrahigh detectivity. Nano Lett 17(1):453–459. https://doi.org/10.1021/acs.nanolett.6b04449

18. Wu Y, Li D, Wu CL, Hwang HY, Cui Y (2023) Electrostatic gating and intercalation in 2D materials. Nat Rev Mater 8(1):41–53. https://doi.org/10.1038/s41578-022-00473-6
19. Zhang K et al (2021) Visualizing Van der Waals epitaxial growth of 2d heterostructures. Adv Mater 33(45):1–10. https://doi.org/10.1002/adma.202105079
20. Pham PV et al (2022) 2D heterostructures for ubiquitous electronics and optoelectronics: principles, opportunities, and challenges. Chem Rev 122(6):6514–6613. https://doi.org/10.1021/acs.chemrev.1c00735
21. Seo J et al (2021) Ultrasensitive photodetection in $MoS_2$ avalanche phototransistors. Adv Sci 8(19):1–9. https://doi.org/10.1002/advs.202102437
22. Mheen B, Shim JS, Kim KS, Oh M, Kwon YH, Nam E (2011) Thee-dimensional eyesafe laser RADAR sytem based on InGaAs/InP 4x4 APD array. In: 2011 3rd International Asia-Pacific conference synthetic aperture radar, APSAR 2011, pp 934–936
23. Jiang Y, Chen J (2019) Optimization of the linearity of InGaAs/InAlAs SAGCM APDs. J Light Technol 37(14):3459–3464. https://doi.org/10.1109/JLT.2019.2917262
24. Zheng J et al (2022) Dynamic-quenching of a single-photon avalanche photodetector using an adaptive resistive switch, pp 1–12. https://doi.org/10.1038/s41467-022-29195-7
25. Zappa F, Lotito A, Giudice AC, Cova S, Ghioni M (2003) Monolithic active-quenching and active-reset circuit for single-photon avalanche detectors. IEEE J Solid-State Circuits 38(7):1298–1301. https://doi.org/10.1109/JSSC.2003.813291
26. Mukhatov A, Le T-A, Pham TT, Do TD (2023) A comprehensive review on magnetic imaging techniques for biomedical applications. Nano Sel. 4(3):213–230. https://doi.org/10.1002/nano.202200219
27. Guo X et al (2016) High-performance graphene photodetector using interfacial gating. Optica 3(10):1066. https://doi.org/10.1364/optica.3.001066
28. Xiang M, Liu F, Liu J, Dong X, Liu Q, Shao X (2024) Computational optical imaging: challenges, opportunities, new trends, and emerging applications. Front Imaging 3. https://doi.org/10.3389/fimag.2024.1336829
29. Abdusalomov AB, Mukhiddinov M, Whangbo TK (2023) Brain tumor detection based on deep learning approaches and magnetic resonance imaging. Cancers (Basel) 15(16). https://doi.org/10.3390/cancers15164172
30. Lozovoy KA et al (2023) Silicon-based avalanche photodiodes: advancements and applications in medical imaging. Nanomaterials 13(23). https://doi.org/10.3390/nano13233078
31. Barbe DF (1976) Charge-coupled device and charge-injection device imaging. IEEE Trans Electron Devices 23(2):177–182. https://doi.org/10.1109/T-ED.1976.18371
32. Wong W, Icel A, Liou JJ (1996) Model for MOS failure prediction due to hot-carriers injection. In: Proceedings IEEE Hong Kong electron devices meeting, pp 72–76. https://doi.org/10.1109/hkedm.1996.566313

# Chapter 4
# Single-Photon Imaging Technology

**Abstract** Single-photon imaging is an advanced method for capturing high-resolution images by detecting individual photons, enabling applications in quantum communication and medical diagnostics. Comprising a transceiver, photon detector, and signal processor, it excels in precise photon detection and signal re-aggregation. Recent advancements have enhanced imaging quality, addressing limitations in human vision. Key applications include quantum information science, bioluminescence imaging, and tumor detection. In particular, laser 3D imaging, bolstered by surface-array single-photon detectors, enables precise 3D reconstruction in low-light environments, benefiting military, medical, and automotive sectors. Challenges like cost, sensitivity, and stability remain, but future developments aim to enhance photon sources and detectors, optimize quantum communication protocols, and integrate advanced algorithms for real-time image processing.

**Keywords** Single-photon imaging · Quantum key distribution · Quantum communication · Photon source · Quantum teleportation · Image processing · Quantum technology · Qubit

## 4.1 Basic Overview of Single-Photon Imaging

1. **Basic concepts of single-photon imaging**

(1) **Definition of single-photon imaging**

Single-photon imaging is an advanced imaging technology that uses individual photons to capture high-definition and high-resolution images. It works by emitting single photons that are directed toward selected points, where they are detected and converted into electrical signals. The system consists of three key components: the transceiver, which transmits and receives the photons; the detector, which converts the photons into electrical signals; and the processor, which processes these signals to produce high-quality images.



S. Du and W. Yin, *Photodetection and Image Sensing Techniques*, Advanced Topics in Science and Technology in China 2, https://doi.org/10.1007/978-981-96-5479-6_4

(2) **Distinguishing single-photon imaging from multi-photon imaging**

Single-photon imaging [1, 2] and multi-photon imaging [3, 4] differ in their mechanisms and applications. Single-photon imaging relies on detecting and re-aggregating individual photons to produce high-definition and high-resolution images. In contrast, multi-photon imaging involves a nonlinear optical process where a material absorbs multiple photons simultaneously, causing a transition from the ground state to the excited state. Multi-photon imaging (e.g., dual- or three-photon imaging) offers advantages over single-photon imaging in terms of tissue penetration, organism protection, SNR, and spatiotemporal resolution. This is because multi-photon imaging uses longer wavelengths and lower energy, improving tissue penetration and reducing organismal damage. Additionally, multi-photon imaging allows for deeper imaging and higher spatial resolution.

2. **Key technologies for single-photon imaging**

(1) **High-efficiency technology of single photon source**

The single-photon source [5] is fundamental to single-photon imaging technology, with its brightness and efficiency directly impacting the sensitivity and speed of imaging. To achieve high efficiency, researchers have employed techniques such as QD and single-molecule fluorescence technologies. These methods enable single-photon sources to produce highly energetic photon output within specific wavelength ranges while maintaining high luminous efficiency and stability.

(2) **Design and optimization of high-performance photon detectors**

High-performance photon detectors are crucial to single-photon imaging technology. To achieve efficient SPD, researchers have focused on optimizing detector design. Enhancements in response speed, sensitivity, and SNR have been achieved by refining material selection, structural design, and process parameters. Additionally, researchers have developed new single-photon imaging methods using advanced quantum statistical techniques and computational imaging, further improving imaging quality and efficiency.

(3) **High-precision scanning and control system**

The high-precision scanning and control system [6] is a crucial component of single-photon imaging technology. It enables accurate scanning and control of the target area, ensuring high-quality image data. To achieve this precision, researchers have utilized advanced control algorithms and precise mechanical structures to maintain stability and accuracy during scanning. Furthermore, the efficiency and accuracy have been enhanced by optimizing scanning speed and resolution.

(4) **Advanced image processing and analysis algorithm**

In single-photon imaging, acquired image data undergoes processing and analysis to produce the final result. Advanced image processing and analysis algorithms are essential for tasks such as filtering, enhancement, edge sharpening, texture analysis, and image segmentation, helping to extract useful information and eliminate noise. Researchers have also developed various feature recognition algorithms, such as character, QR code, and face recognition, to address different application needs.

3. **Significance and role of optoelectronic imaging technology**
(1) **Limitations of human eye vision**

The human eye is a highly sophisticated and functional imaging system, working in conjunction with the brain's optic nervous system. It allows us to perceive the external environment, assess target distance, size, details, and gray levels, and adapt to changes in lighting conditions. This unique visual ability helps humans understand the world and acquire knowledge, working alongside other sensory organs such as the ears (hearing), nose (smell), tongue (taste), and skin (touch). In the exchange of subjective and objective information, more than 80% of sensory input comes from visual information obtained through the naked eye (without the aid of optical equipment). However, human vision has four key limitations [7]:

1. Spectral range is limited: only visible light (0.38–0.76 μm) can be distinguished.
2. Limited response speed: it can only distinguish between stationary or slow-moving targets, with a time resolution of tens of milliseconds.
3. The dynamic range of the human eye spans about 1 million to 1 billion times in light intensity, allowing it to perceive both very dim and very bright environments. It can detect light as faint as $10^{-6}$ lux (starlight) and handle brightness up to $10^{3}$ lux (direct sunlight).
4. Limited spatial resolution: The spatial resolution of the human eye is around 1 arcminute of detail in ideal conditions, corresponding to a resolution similar to a 576-megapixel camera for the central visual field.

(2) **Functionalities and technical characteristics of the optoelectronic imaging system**

Humans have invented various optical and optoelectronic imaging systems to make up for the above inherent defects of human eyes, improve their spatial and temporal resolution capabilities, and broaden the spectral response range and the dynamic range. These optical and optoelectronic instruments include:

1. Optical imaging instruments: microscope, telescopes cameras, and all kinds of optoelectronic imaging systems.
2. Spectral analysis instrument: prism/grating spectral instrument, Fourier transform spectrometer, etc.
3. Optical metering instruments: length, angle, flatness, surface roughness, optics, luminosity, and spectral metering equipment.
4. Optical test instruments: measuring microscope, tool microscope, side goniometer, theodolite, quadrant instrument, level instrument, etc.

Optoelectronics is a specialized discipline focused on the physical processes involving photons, electrons, and electron–hole pair interactions, such as their conversion, amplification, processing, transmission, reception, and display. A key component of optoelectronics is modern optoelectronic imaging (PEI) technology, which uses advanced imaging devices and systems designed for high sensitivity, wide spectral range, fast response, and large dynamic range. This technology compensates for the human eye's limitations in spatial, temporal, and spectral sensitivity. It enables the visualization of static and dynamic scenes involving faint light, infrared, ultraviolet, X-rays, gamma rays, and other forms of electromagnetic radiation that are otherwise invisible or difficult to perceive. Optoelectronic imaging has become a vital high-tech tool, providing over 80% of external information for applications in observation, targeting, range finding, tracking, guidance, and alarm systems. It is also widely utilized in aerospace remote sensing, high-speed photography, low-light detection, medical diagnostics, and biological research.

Optoelectronic imaging systems can be classified based on their spectral response band or intended purpose. Spectrally, they include γ-ray, X-ray, ultraviolet, visible light, infrared, and microwave imaging systems [1–7]. Functionally, they serve as observation, aiming, ranging, tracking, guidance, alarm, and diagnostic systems. Acting as the "eyes" of various equipment, these systems capture, process, and analyze target image data, enabling critical tasks such as early threat detection and neutralization in military operations. In civilian applications, they support industrial measurement and control, agricultural forecasting, scientific research, public security, and medical diagnostics. To perform these roles effectively, optoelectronic imaging systems require broad spectral responsiveness (from γ-rays to microwaves), a large dynamic range ($10^{-5}$ lx to $10^{5}$ lx), fast response times (from milliseconds to femtoseconds), and the ability to transmit over ultra-long distances. Additionally, they must balance reliability, compactness, affordability, and performance to meet technical and practical demands as depicted in Fig. 4.1.

### (3) Composition and working principles of optoelectronic imaging system

As illustrated in Fig. 4.2, an optoelectronic imaging system typically comprises several components: the observed target, a transmission medium, an optical imaging objective lens, an optoelectronic imaging device, a power supply and electronic processing unit, and a terminal display. The imaging process involves radiation or reflection from the observed scene passing through a transmission medium, which may cause some attenuation. The optical objective lens focuses this radiation onto the optoelectronic imaging device's surface. Through the device's unique photoelectric

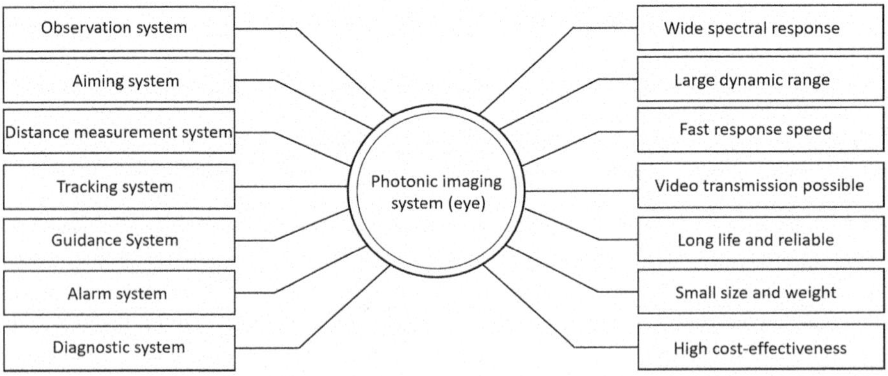


**Fig. 4.1** Applications, functions and technical characteristics of the optoelectronic imaging system

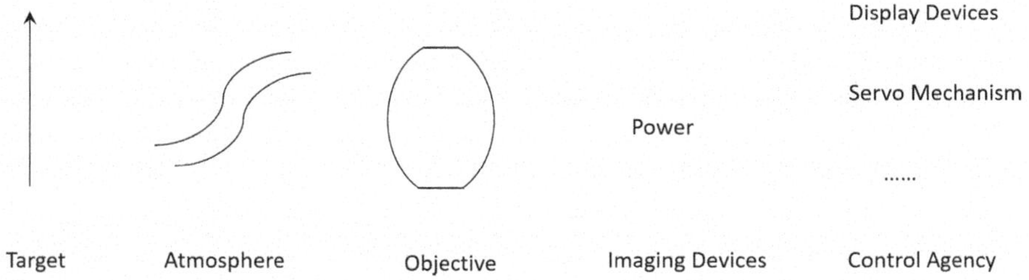


**Fig. 4.2** Schematic diagram of the composition and working principles of the optoelectronic imaging system

conversion, amplification, processing, and display capabilities, the system transforms electromagnetic radiation—including weak light, infrared, ultraviolet, X-rays, and gamma rays—into visible light images, making otherwise imperceptible static and dynamic scenes visible.

(4) **Types and working principles of optoelectronic imaging devices**

The optoelectronic imaging device is the core component of an optoelectronic imaging system. It plays a crucial role in enabling the system to detect, identify, and classify specific targets, and serves as the foundation for designing other system components, such as the optical lens unit and electronic image processing unit. In summary, the device functions through mechanisms of photoelectric conversion, amplification, processing, and display, which work together to enable the system's overall performance.

1. A **solid-state optoelectronic imaging device** can take the form of a unit detector, line array detector, or surface array detector. Its working principle is based on the photoelectric effect within a semiconductor, specifically using the photovoltage or photoconductive effects. During operation, the device converts the radiation into a charge signal that is proportional to the number of input photons at each pixel. The device then processes these signals through internal and external drive signals, outputting them as video signals via a TV system with frame scanning. These video signals can be used for display observation, tracking, guidance, or transmitted as radio signals through an antenna for remote analysis and reference. Devices in this category include infrared detectors (for medium-wave and long-wave infrared), CCDs in digital cameras, CMOS-CCD (integrating optical array circuits and AD circuits in one chip), and EMCCD (low-light CCDs) with an avalanche electron multiplication mechanism [8–11].
2. **Vacuum optoelectronic imaging devices** typically feature an ultra-high vacuum image enhancer design [12]. This setup includes a time pole, a microchannel plate (MCP), a fluorescent screen, an electronic optical lens, and an integrated high-voltage power supply. The device operates based on the semiconductor photoelectric effect, utilizing varying spectral responses to convert incident light or reflected images into a vacuum optoelectronic image. The input radiation is

converted into proportional pixel values and photon counts. The energy is then focused through an optical lens, which bombards the microchannel plate for electron multiplication, and the final step involves converting the electrons back into visible light. This enhanced light image, which can be amplified by a factor of $10^4$, is either directly observed or coupled to a CCD for TV display. Various types of micro-light image enhancers are available, including those for X-rays, UV, and near-infrared, as well as high-speed photographic tubes. Micro-light video devices include technologies like ICCD (intensified CCD), and high-frequency, low-noise photon counters.

3. **Image display devices** include various technologies such as electroluminescent screens, molecular orientation displays, plasma display panels, semiconductor LED array displays, organic semiconductor LED array displays, and X-ray fluorescence conversion screens. These devices are used to provide visible images for medical diagnosis and deep-space short-wave radiation detection systems. Additionally, there are devices like computer scanners and laser printers, which use laser beams modulated by image signal sound and light, as well as large projection display systems.

4. **Applications of single-photon imaging**

(1) **Scientific research field**

A. **Quantum information science**

Single-photon imaging is essential in quantum information science, a rapidly advancing field that leverages quantum mechanics to process and transmit information. This technology offers high-resolution and sensitive imaging, which is crucial for quantum research. In quantum communication, single-photon imaging is used to detect and measure single photons, enabling the transmission and entanglement of QBs. It allows precise measurement of QB states and entanglement, enhancing the reliability and security of quantum communication. Additionally, single-photon imaging plays a key role in quantum computing and quantum sensing. In computing, it aids in observing and manipulating individual QBs, advancing quantum computers. In quantum sensing, it provides precise measurements of physical quantities like magnetic fields, temperature, and pressure.

B. **Bioluminescence imaging**

Bioluminescence imaging is a key biological technique that uses fluorescence signals emitted by fluorescent materials when exposed to light at specific wavelengths to observe and analyze biological samples. Single-photon imaging plays a crucial role in bioluminescence imaging by enabling the accurate detection and imaging of individual fluorescent molecules. This highly sensitive technique allows scientists to study interactions and dynamic processes between biomolecules, providing insights into the structure and function of biological systems. In single-molecule imaging, single-photon technology offers significant advantages, delivering high-resolution images that reveal the distribution and behavior of biomolecules within cells. This is essential for studying biological processes like protein interactions, DNA replication, and RNA transcription.

C. **Nanoscale imaging**

Single-photon imaging techniques are widely used in nanoscale imaging, which involves observing and analyzing matter at the nanometer scale. This technology provides high-resolution, sensitive imaging, making it essential for nanoscale studies. In nanophotonics, single-photon imaging is employed to examine light-matter interactions at the nanoscale. It allows scientists to observe the optical properties of individual nanoparticles and nanostructures, deepening the understanding of optical phenomena at this scale. Additionally, single-photon imaging plays a key role in nanoelectronics and nanomechanics, enabling the observation and manipulation of individual devices and structures, thus advancing the development of these fields.

(2) **Medical field**

A. **Fluorescence-labeled imaging**

Fluorescence-labeled imaging is a key application of single-photon imaging in medicine. This technique involves imaging samples labeled with biomolecules, such as fluorescent proteins or dyes. Fluorescence-labeled imaging offers high sensitivity, resolution, and contrast, providing detailed information about cell structures and tissues. In medicine, it is widely used in oncology, neuroscience, cardiovascular research, and more. With this technique, doctors and researchers can observe tumor growth, metastatic processes, and neuron interactions in real-time. Additionally, fluorescence-labeled imaging is valuable for tracking drug molecule distribution and mechanisms, offering crucial insights for drug development and clinical applications.

B. **Tumor detection and diagnosis**

Single-photon imaging technology offers significant advantages in tumor detection and diagnosis [13]. It allows doctors to observe real-time blood flow, metabolic activity, and cell proliferation in tumor tissues, which are crucial for assessing tumor malignancy, evaluating treatment effectiveness, and predicting prognosis. Additionally, single-photon imaging can be combined with fluorescence labeling techniques for targeted imaging of tumor-associated molecules. For example, using fluorescent-labeled antibodies and tumor-associated antigens enables clear visualization of tumor location and distribution. This enhances early detection, precise staging, and the formulation of treatment plans for tumors.

C. **Neural activity monitoring**

Neural activity monitoring is another key application of single-photon imaging in medicine. This technology allows researchers and doctors to observe real-time interactions between neurons, neurotransmitter release, and signaling processes, making it valuable for studying neurological disorders, psychosis, and cognitive impairments. By combining single-photon imaging with fluorescent labeling techniques, researchers can monitor neuronal activity and neurotransmitter release. For example, fluorescently labeled neurotransmitter receptor antibodies enable the observation of neurotransmitter transmission between neurons. Additionally, single-photon imaging is useful for studying changes in neuronal morphology and synaptic structures, supporting neuroscience research.

(3) **Security field**

A. **Cryptography**

Traditional communication methods, such as radio waves or optical fibers, are vulnerable to interference and eavesdropping. Single-photon imaging enables quantum communication, where information is transmitted using the quantum state of a single photon, ensuring high security and confidentiality. This encrypted communication often employs QKD, where the parties first share a random key to encrypt and decrypt information. Since the quantum state of a single photon is altered by any external observation or interference, any such attempt can be detected by the communicating parties. This ensures that the transmission remains secure, preventing interception or tampering by third parties.

B. **Lidar detection**

Lidar, which uses lasers for ranging and detection, is widely used in military, aerospace, and geological exploration. The integration of single-photon imaging significantly enhances the accuracy and efficiency of lidar systems. It enables high-resolution imaging of targets, improving the ability to identify and classify them. Additionally, single-photon imaging allows for real-time monitoring and tracking of targets, enhancing the reaction speed and operational efficiency of lidar systems.

C. **Identification of hidden targets**

In military, anti-terrorism, and reconnaissance operations, identifying and locating hidden targets is often crucial. Single-photon imaging technology enables target imaging in low-light or no-light conditions, facilitating identification and positioning in environments with low visibility or darkness. Additionally, it can penetrate obstacles such as smoke, dust, or camouflage, allowing for the accurate identification and location of targets, even if they are intentionally or unintentionally concealed.

5. **The challenges of single-photon imaging**

(1) **Technical bottlenecks and cost issues**

While single-photon imaging technology offers many advantages, it faces several technical and cost challenges in practical applications. First, it requires high-precision optical systems and electronic devices, which are expensive to produce and require regular maintenance and calibration, increasing costs. Additionally, single-photon imaging relies on special radioactive substances that are costly to produce and can contribute to environmental pollution. Furthermore, the technology faces technical bottlenecks in image processing and analysis, necessitating the development of new algorithms and software to enhance imaging quality and efficiency.

(2) **Security and privacy protection**

Security and privacy protection are critical concerns in the practical application of single-photon imaging technology. Due to its high spatial resolution and ability to penetrate materials, it could potentially be used to violate personal privacy or conduct illegal surveillance. Therefore, it is essential to establish safety regulations and standards to ensure the legal, safe, and ethical use of this technology. Additionally, new privacy protection measures, such as image encryption techniques, must be developed to safeguard personal privacy and data security.

## 4.2 Laser 3D Imaging Based on a Surface-Array Single-Photon Detector

1. **The development status of laser 3D imaging technology**

Laser 3D imaging [14–17] has made significant advancements, showcasing diverse applications across various fields. Compared to traditional 2D remote sensing, laser 3D imaging provides precise 3D information about targets, enabling accurate identification through software processing. This technology offers distinct advantages for both military and civilian uses. Globally, research in laser 3D imaging has matured. The development began with MIT's Lincoln Laboratory, which introduced the first coherent $CO_2$ lidar in the 1960s, sparking worldwide advancements. Companies like Raytheon have since achieved high-resolution lidar imaging systems. China has also made notable contributions to this field. For example, Professor Hu Yihua's team at the National University of Defense Technology developed a reflective tomographic lidar system, achieving 3D super-resolution reconstruction of targets over distances of 10.38 km. This innovation supports ultra-long-distance imaging of small objects. Laser 3D imaging technology spans military, medical, and scientific domains. In medical imaging, it enables clear visualization of organs and lesions, improving diagnostic accuracy. In cell biology and neuroscience, it facilitates detailed observation of cellular structures, neurons, and synapses, advancing the understanding of life processes. In oncology, the technology aids in analyzing cancer cells' 3D structures, providing insights into tumor growth and morphology.

2. **Basic principles of laser 3D imaging technology**

(1) **Working mechanism of laser imaging**

Laser imaging involves using a laser beam to scan an object. When the laser beam strikes the object's surface, it interacts with the surface and produces a reflection. This reflected beam is captured by a receiver and converted into an electrical signal, forming an image of the object. Owing to its high degree of focus, laser imaging can generate high-quality images. A typical laser imaging system comprises a laser, lens, scanning device, and receiver. The laser generates the beam, the lens focuses it, the scanning device controls the beam's movement across the object's surface, and the receiver collects the reflected signal to convert it into an image.

(2) **Operational mechanism of 3D imaging**

3D imaging is to convert real-world objects or scenes into 2D images or videos while retaining the objects' 3D information. Common methods for acquiring 3D data include lidar, stereo cameras, structured light, and motion-sensing devices. During the 3D imaging process, data is collected on the shape, size, position, and other attributes of the object. This data undergoes processing steps such as denoising, filtering, and registration to enhance quality. The refined data is then used to create a 3D model through 3D reconstruction techniques. Finally, the 3D models can be displayed on a computer screen, printed into physical images, or presented in other formats.

(3) **Operational mechanism of surface array single-photon detector**

A surface array single-photon detector is a device designed to detect single-photon events. Its basic principle involves a change in the number of charge carriers within the detector when a photon passes through it. This change is detected by the circuit within the detector and processed through a signal processing module to confirm the presence of photons. A surface array single-photon detector typically comprises multiple SPD units arranged in a 2D array. Each unit independently detects photon events and sends the results to the signal processing module for further analysis. This design allows the detector to capture the 2D spatial distribution of photon events. In a laser 3D imaging system, a surface array single-photon detector is used to receive reflected laser signals and convert them into electrical signals, forming a 3D image of the object. With its high sensitivity, excellent time resolution, and low noise, the face array single-photon detector significantly enhances the performance of laser 3D imaging systems.

3. **Laser 3D imaging technology based on surface array single-photon detector**

(1) **Laser 3D imaging based on a surface array single-photon detector**

Laser 3D imaging technology based on single-photon detectors uses these detectors to probe laser pulses and achieve 3D contour reconstruction of an object's surface by analyzing the number and timing of the received photons. This technology offers high resolution, sensitivity, and real-time performance, making it widely applicable in fields like industrial inspection, robot navigation, and biomedicine. A laser pulse is generated with adjustable parameters, such as frequency, wavelength, and energy, depending on the specific application requirements. When the laser pulse irradiates the object's surface, some photons are reflected, refracted, or absorbed, leading to scattering. These scattered photons are collected by an optical system, focused, and directed onto a single-photon detector. The surface array single-photon detector can simultaneously detect multiple photons and record their arrival times and intensity. Using the photon data from the detector, the 3D coordinates of the object's surface are calculated through specific algorithms and mathematical models, enabling the reconstruction of the 3D contours of the object. This laser 3D imaging technology, using a surface array single-photon detector, is known for its high resolution, sensitivity, and real-time capabilities, making it ideal for accurate measurement and surface reconstruction. Additionally, the technology can be adapted to different application scenarios by adjusting the laser pulse parameters.

(2) **Composition of a laser 3D imaging system based on a surface array single-photon detector**

A. **Laser emission module**

The laser emission module is a crucial component of the laser 3D imaging system, responsible for emitting the laser beam. This module typically consists of a laser, laser driver, beam extender, beam shaper, beam finder, and beam pointing controller. The laser generates the beam, and the laser driver controls the power and pulse frequency of the beam. The beam extender and beam shaper are used to enlarge and modify the laser beam, enabling it to cover a larger scanning area. The beam pointing controller directs the laser beam, ensuring it accurately scans the target object.

B. **Optical system**

The optical system is responsible for focusing the laser beam onto the target object and collecting the reflected or scattered laser signals. The system typically includes an objective lens, beam extender, beam shaper, beam splitter, and lens group. The objective lens focuses the laser beam on the target, while the beam extender and beam shaper enlarge and shape the beam. The beam splitter divides the laser beam into two parts: one illuminates the object, and the other captures the reflection or scattering from the object. The lens group collects and focuses the reflected or scattered laser signals for further processing.

C. **Surface-array single-photon detector**

It is responsible for receiving laser signals reflected or scattered by objects and converting them into electrical signals. The detector typically consists of a photon detector, signal amplifier, and signal processor. The photon detector detects individual photon events, while the signal amplifier boosts the weak output signals from the detector. The signal processor then analyzes and processes the amplified signals to generate the final data for the imaging system.

D. **Signal processing and data acquisition unit**

The signal processing and data acquisition unit is a crucial component of the laser 3D imaging system, responsible for processing and collecting the signals output by the detector. This unit typically includes a signal processor, data collector, and digital/analog converter. The signal processor filters, amplifies, and shapes the signal from the detector to enhance its quality. The data collector gathers the processed signal and stores it, while the digital/analog converter converts analog signals into digital form for computer processing.

E. **Control system**

This system is responsible for managing and adjusting the overall operation of the system. It typically includes a controller, actuator, and sensor. The controller receives processing instructions from the computer, resolves them, and executes the necessary actions. The actuator adjusts the system based on the controller's instructions, such as controlling the direction of the laser beam or regulating the detector's exposure time. The sensor monitors the system's operating status and provides feedback to the controller, ensuring proper functioning.

F. **Computer processing and 3D reconstruction unit**

The computer processing and 3D reconstruction unit is the final step in the laser 3D imaging system, responsible for processing and analyzing the collected data to generate a 3D image of the target object. This unit typically consists of a computer, 3D reconstruction algorithm, and display. The computer processes and analyzes the collected data, while the 3D reconstruction algorithm reconstructs the target object's 3D image from the data. The display is used to show the reconstructed 3D images for visualization and further analysis.

(3) **Advantages of laser 3D imaging technology based on surface array single-photon detector**

The surface array single-photon detector operates effectively under very low light conditions, enabling accurate imaging even in weak light environments. This capability is particularly valuable for applications such as night navigation and astronomical observation. It can precisely capture the arrival time of photons, providing a significant advantage in obtaining depth information, making it ideal for tasks like deep learning and 3D reconstruction. While imaging techniques are often vulnerable to interference from ambient and scattered light, the face-array single-photon detector effectively suppresses these interferences, maintaining high imaging quality in complex environments.

Moreover, surface-array single-photon detectors allow for high-speed data acquisition, which is crucial for real-time 3D imaging of dynamic scenes. This feature is especially important in applications such as robotic navigation and autonomous driving, where real-time 3D information about the surroundings is critical. Laser 3D imaging technology based on surface array single-photon detectors is also easily miniaturized and integrated, making it suitable for portable devices and compact systems. In addition to 3D imaging, this technology can be configured for other functions, such as laser ranging and velocity measurement, by adjusting the detector and light source setup. With its high sensitivity, resolution, anti-interference capabilities, fast imaging, miniaturization potential, and multifunctionality, laser 3D imaging technology based on surface array single-photon detectors has broad potential across many fields.

4. **Key problems of laser 3D imaging technology based on surface array single-photon detector**

In the process of implementing the laser 3D imaging technology, there are several key problems to be solved and optimized:

(1) **Performance optimization of surface array SPD**

A. **Improving sensitivity**

Sensitivity refers to the detector's ability to respond to photon signals. Increasing sensitivity means the detector can more effectively detect weak photon signals, enhancing the overall detection capability of the imaging system. Solutions to improve sensitivity may include using new photosensitive materials, optimizing the photoelectric conversion structure of the detector, and upgrading read-out circuits and signal processing algorithms. These improvements can enhance the detector's response efficiency and SNR, leading to better detection and imaging performance.

B. **Improving temporal resolution**

Temporal resolution refers to the detector's ability to precisely resolve the arrival time of photons. In laser 3D imaging, accurately measuring the photon's flight time is crucial for achieving high-precision distance measurements. To improve temporal resolution, optimizations can be made to the detector's readout circuit and signal processing algorithms, reducing time jitter and latency. Additionally, the use of higher-performance electronic components and faster signal processing speeds are effective strategies for enhancing temporal resolution.

C. **Reducing dark count rate**

The DCR refers to the proportion of mis-triggered events generated by the detector in the absence of light signals. A high DCR can degrade the SNR and reduce the detection efficiency of the imaging system. To lower the DCR, strategies include optimizing the detector's material and structure, lowering the detector's temperature, and enhancing the readout circuit and signal processing algorithms. These measures help mitigate the effects of internal noise and external interference, thereby improving imaging quality.

(2) **The matching problem between the laser light source and the detector**

A. **Light source wavelength selection**

The selection of the source wavelength is a crucial consideration. First, the wavelength of the light source affects the detection efficiency of the detector. For surface array single-photon detectors, choosing a light source wavelength that matches the detector's response band enhances detection sensitivity. Secondly, the wavelength influences the propagation characteristics of the laser in the medium, including factors such as scattering and absorption, which can impact imaging quality. Therefore, in practical applications, it is essential to select the appropriate light source wavelength based on the specific imaging requirements and the performance characteristics of the detector.

B. **Power and pulse width of light source**

The power of the light source directly impacts the SNR of the detector. Within a certain range, increasing the power can boost the intensity of the detection signal, improving the SNR and enhancing imaging quality. However, excessive power may cause thermal damage to the detector, so it is essential to control the light source power carefully. Furthermore, the pulse width of the light source affects both the laser's propagation distance and the imaging speed in the medium. Generally, a shorter pulse width leads to faster imaging speeds, but it also increases scattering and absorption, which can degrade imaging quality. Therefore, it is necessary to balance the power and pulse width of the light source based on the specific application requirements.

C. **Temporal synchronization of the light source and the detector**

Temporal synchronization between the light source and the detector is essential for achieving high-precision 3D imaging. Because of the finite speed of light, there is a delay for the laser to reach the detector. If the light source is not synchronized with the detector's timing, it can distort the signal received, negatively impacting the imaging results. Therefore, precise temporal control and synchronization techniques are required to ensure the light source and detector are properly aligned in time, enabling accurate and high-precision 3D imaging.

(3) **Stability and reliability of the imaging system**

The stability and reliability of the system are the key to long-term stable operation and wide application. In laser 3D imaging systems, stability and reliability are influenced by several factors.

A. **Environmental adaptability**

Environmental adaptation refers to the ability of the imaging system to operate stably under various environmental conditions. Since the laser 3D imaging system may be deployed in complex environments, such as those with temperature fluctuations, humidity changes, and vibration interference, it must be designed to withstand these factors. Solutions include using components with a wide temperature operating range, designing effective heat dissipation and dust protection structures, and ensuring the system has seismic resilience. Additionally, software algorithms can be employed to compensate for environmental changes, further enhancing the system's adaptability.

B. **Reliability**

Reliability is crucial for ensuring the long-term, stable operation of the imaging system. It encompasses hardware design, software design, power management, and other system aspects. In hardware design, redundant and fault-tolerant technologies can be implemented to enhance reliability and the system's ability to handle failures. For software design, modular design, exception handling mechanisms, and other techniques can help reduce software failure rates and improve system stability. Additionally, effective power management and heat dissipation design are vital for enhancing reliability. By optimizing power management strategies, the system can maintain stable operation under varying load conditions, while efficient heat dissipation reduces system temperature and extends component lifespan.

C. **System calibration and maintenance**

System calibration and maintenance [18] are essential for maintaining the performance of imaging systems. Due to potential errors and drifts in the system, regular calibration and maintenance are required. Calibration includes optical system and detector calibration, aimed at eliminating systematic errors and enhancing measurement accuracy. Maintenance involves tasks such as cleaning optical components, checking circuit connections, and updating software versions to ensure the system operates normally and prolong its service life. By establishing a robust calibration and maintenance system, issues can be identified and addressed promptly, ensuring the stability and reliability of the imaging system. Additionally, this approach improves the system's maintainability and scalability, making future upgrades and enhancements easier.

5. **The solutions of laser 3D imaging technology based on surface array single-photon detector**

(1) **Improving sensitivity and resolution of the surface array single-photon detector**

A. **Selecting high-performance materials**

The sensitivity and resolution of face array single-photon detectors are directly influenced by the material properties used. To enhance detector performance, high-purity semiconductor materials, such as InGaAs, can be selected. These materials offer a high optical absorption coefficient and excellent electron mobility, which improve the detector's sensitivity and resolution. Additionally, the material's band structure can be adjusted through doping with different elements to optimize it for specific laser 3D imaging applications. For instance, doping with phosphorus (P) can enhance the response wavelength of InGaAs, enabling it to detect shorter wavelength laser light.

B. **Optimizing detector structure**

The sensitivity and resolution of the detector can be further improved by optimizing its structure. One common approach is to incorporate a microcavity structure, which increases the photon interaction time inside the detector, thereby enhancing PDE. Additionally, a multilayer structure can be used to combine the photon detection layer with a signal amplification layer, improving the SNR. For example, a layer of InGaAsP can be grown on the InGaAs layer to serve as a signal amplification layer, further enhancing the detector's sensitivity.

C. **Improving the readout circuit**

The performance of the readout circuit is crucial for the sensitivity and resolution of the face array single-photon detector. To enhance detector performance, a high-speed, low-noise readout circuit can be employed. Additionally, a multi-channel readout circuit can be used to process signals from multiple detectors in parallel, thereby increasing overall data processing speed. Furthermore, optimizing the noise performance of the readout circuit can further improve the detector's resolution.

(2) **Optimize the matching between the laser light source and the detector**

A. **Selection of the appropriate laser wavelength**

The laser wavelength should align with the surface properties of the object being measured to optimize absorption and reflection. For instance, near-infrared lasers are more suitable for black or dark objects, while white or light objects may require different wavelengths. Second, the laser wavelength must match the sensitivity of the single-photon detector, as detectors typically respond more efficiently to specific wavelengths. Selecting a wavelength that aligns with the detector's sensitivity can enhance detection efficiency and imaging quality. Finally, the wavelength choice must also consider the system's resolution and depth range. Generally, shorter wavelengths provide higher resolution but a reduced depth range, while longer wavelengths offer greater depth range but lower resolution. Therefore, a balance between resolution and depth range must be struck when selecting the laser wavelength.

B. **Manipulating laser power and pulse width**

The laser power must be sufficient to generate enough reflected and scattered light from the object's surface to be detected by a single-photon detector. However, excessive laser power may damage the object's surface, so it should be adjusted carefully to ensure imaging quality while avoiding thermal damage. The laser pulse width directly impacts the system's resolution and depth range. Typically, shorter pulse widths offer higher resolution but reduce the depth range, while longer pulse widths provide a greater depth range at the cost of lower resolution. Therefore, a balance between resolution and depth range must be considered when adjusting the pulse width. Additionally, the detector's response speed plays a role in pulse width adjustment; if the detector is too slow, a short pulse width may lead to incomplete or inaccurate signal detection, negatively affecting the imaging quality.

C. **Ensuring time synchronization**

Temporal synchronization is a critical aspect of laser 3D imaging, ensuring that the emission of each laser pulse and the detector's response occur within the same time window. This precise control over the timing of both the laser source and the single-photon detector guarantees that each laser pulse is effectively detected, thereby improving the imaging quality and system stability. Several methods can be used for achieving time synchronization, such as employing synchronization clocks and compensating for fiber transmission delays. In practical applications, selecting the most suitable time synchronization method depends on the specific requirements and conditions of the system. Additionally, the synchronization mechanism must be carefully calibrated to ensure reliable and stable synchronization across various operating environments.

(3) **Improving design and control of the imaging system**

A. **Optimize the optical system**

To enhance the performance of the laser 3D imaging system, high-quality optical components should be selected to minimize the effects of stray light and thermal noise. This ensures that the detected signal is as accurate as possible. Additionally, advanced coating technologies can be applied to the optical elements, reducing light reflection and scattering, which further improves the overall image quality. Furthermore, adjusting the focal length and optical path of the system can optimize the resolution and depth range of the imaging system, enabling it to capture more detailed and accurate 3D data of the target objects.

B. **Improving the scanning strategy**

The choice and optimization of the scanning strategy are crucial for achieving high-quality and fast imaging. To enhance scanning efficiency, more advanced algorithms like phase coding or frequency modulation can be adopted, which accelerate the scanning process while maintaining high resolution. Additionally, by introducing multi-angle scanning techniques, the interaction between the laser beam and the object surface can be enhanced. This approach improves the imaging quality by capturing more detailed data and increases the system's ability to resist external interference.

C. **Ensuring system calibration and maintenance**

To ensure the stability and reliability of the laser 3D imaging system, regular calibration and maintenance are essential. Calibration should be performed periodically to correct errors in the optical components and scanning mechanisms, ensuring the accuracy of the system. Additionally, maintaining optimal temperature and humidity control is crucial to minimize the impact of environmental factors on the system's performance. Regular inspections and maintenance help identify potential issues early, ensuring smooth operation and prolonging the system's service life. By following these practices, the overall reliability and longevity of the laser 3D imaging system can be significantly improved.

6. **Application of laser 3D imaging technology for surface array single photon detector**

(1) **Unmanned driving and robot navigation**

The application of laser 3D imaging with surface array single-photon detectors in unmanned driving and robot navigation is highly impactful. The technology provides precise depth information and environment perception, which is essential for autonomous vehicles and robots to navigate and avoid obstacles in complex environments. By utilizing surface-array single-photon detectors, driverless cars and robots can achieve high-precision 3D reconstruction and positioning, enhancing safety and operational efficiency. For unmanned driving, the system enables advanced lidar capabilities that generate 3D maps and offer real-time environmental awareness [19–21]. This allows autonomous vehicles to make informed decisions for navigation and obstacle avoidance, even in challenging conditions like low visibility or fog. The surface array single-photon detector's ability to detect single photons and measure distance with high precision ensures that the vehicle can detect objects at various distances and shapes, ensuring a safe driving experience. In robot navigation, the same technology provides real-time 3D sensing, allowing robots to accurately perceive their surroundings. This enables robots to plan paths and perform tasks in environments like factories, homes, or hospitals, where precise operations are needed. By integrating surface-array single-photon detectors, robots can make intelligent decisions for obstacle avoidance, task execution, and even interaction with humans, advancing automation in various industries.

(2) **Security monitoring and obstacle detection**

This technology is critical in various security and obstacle detection applications, offering enhanced performance and accuracy. In security monitoring, the face array single-photon detector can be used to achieve high-precision 3D face recognition systems. By capturing detailed 3D facial features and combining them with deep learning and AI algorithms, it can offer more accurate and reliable identification compared to traditional 2D recognition methods. This significantly enhances the performance and security of authentication systems, especially in high-security environments like airports, banks, and government facilities. Additionally, the surface array single-photon detector is highly effective for panoramic imaging and comprehensive monitoring of large areas, such as public spaces and critical infrastructure. The 3D imaging capability enables real-time monitoring with a high level of detail, improving the detection of suspicious activity and the overall security response. This technology can track objects and individuals with precise depth information, even in low light or challenging conditions, making it ideal for enhancing surveillance systems. In the realm of obstacle detection, surface-array single-photon detector-based laser 3D imaging systems provide highly accurate obstacle detection for unmanned vehicles and robotics. By measuring the distance and shape of obstacles in real time, the system enables efficient collision avoidance, significantly improving the safety of autonomous systems. This is particularly important for applications in dynamic and complex environments, such as warehouses, urban areas, and factories. For autonomous vehicles, real-time 3D obstacle detection using this technology allows for intelligent driving and better decision-making. It provides precise information about the road environment, including roadblocks, pedestrians, and other vehicles, helping the vehicle safely navigate and avoid collisions. The depth-sensing capabilities of the surface-array single-photon detector also help improve the vehicle's ability to understand its surroundings, even in low visibility conditions such as fog, rain, or nighttime driving.

(3) **Industrial testing and measurement**

This technology is increasingly being applied in the field of industrial detection and measurement. This technology enables high-precision 3D imaging and measurement, which enhances industrial production efficiency and quality control. In industrial detection, surface array single-photon detectors are used for 3D scanning and reconstruction to measure the shape and size of products with exceptional accuracy. Using a laser scanner along with the surface array single-photon detector, high-precision 3D scanning can be achieved to obtain detailed geometric information about objects. This allows industrial engineers to perform comprehensive quality control and defect detection, ensuring products meet the required standards for quality and reliability. Furthermore, surface array single-photon detector-based laser 3D imaging technology is instrumental in industrial measurement and monitoring. By capturing real-time 3D information of products on the production line, the technology enables continuous monitoring and measurement, allowing issues in the production process to be identified and corrected quickly. This capability facilitates a more intelligent and automated industrial production environment, enhancing production efficiency and reducing costs.

## 4.3 Application of Single-Photon Imaging in Low-Light Detection, Quantum Communication and Other Fields

1. **Application of single-photon imaging in low-light detection**

(1) **The importance and challenges of low-light detection**

Low-light detection technology is vital across various fields. In scientific research, it plays a key role in exploring the mysteries of the universe. For example, in astronomy, it enables the observation of distant cosmic phenomena like galaxies and black holes. In the biomedical field, low-light detection is widely used in fluorescence microscopy, allowing scientists to examine the structures and functions of living organisms at a microscopic level. Additionally, it is essential in security, night vision, and optical communication technologies. However, low-light detection faces several challenges. The weak signals are easily disrupted by noise, necessitating highly sensitive detectors. Furthermore, acquiring and processing these signals requires extreme precision to avoid errors. Environmental factors such as temperature, humidity, and electromagnetic interference also complicate the process. To overcome these challenges, single-photon imaging technology has been developed. By detecting the energy of individual photons, this technology offers exceptional sensitivity and accuracy, enabling the effective capture of low-light signals and high-quality optical images.

(2) **Advantages of single-photon imaging technology**

A. **High sensitivity**

Single-photon imaging is highly sensitive due to its ability to detect the energy of individual photons. Unlike traditional imaging techniques, which require millions of photons per pixel to capture even dark images due to the limitations of PMTs, single-photon imaging overcomes this by counting only the photons detected within a specific time window. This allows it to detect light at the single-photon level. As a result, it can produce clear images even in extremely low-light conditions, making it highly valuable for applications in areas like astronomical observation and biological imaging.

B. **High resolution**

Single-photon imaging achieves high resolution by leveraging the wave-particle duality of light. In quantum imaging, photons are detected individually, and by accumulating enough photons, precise images of objects can be reconstructed. This ability to detect single photons allows single-photon imaging to surpass traditional imaging techniques, making it especially valuable for applications requiring detailed images, such as quantum optical microscopy and quantum computing.

C. **Low noise**

In traditional imaging systems, noise primarily arises from sources like thermal and quantization noise. However, single-photon imaging systems are less affected by these noise factors because they rely on detecting individual photons. This is particularly advantageous in low-temperature environments, where thermal noise is greatly reduced. As a result, single-photon imaging systems can produce clearer, less noisy images, which is highly beneficial for applications in scientific research, military reconnaissance, and security monitoring.

2. **Application of single-photon imaging in quantum communication**
(1) **Basic principles and advantages of quantum communication**

Quantum communication leverages quantum superposition states and entanglement effects for the transfer of information. This is based on the fundamental principles of quantum mechanics, including uncertainty, measurement collapse, and the no-cloning theorem, which ensure that information cannot be intercepted, copied, or predicted during transmission. These principles provide an unprecedented level of security in communication, guaranteeing that transmitted information remains secure against eavesdropping and computational attacks.

Quantum communication also offers the potential for higher transmission rates compared to classical communication systems. This is because quantum systems can transmit large amounts of information rapidly, making quantum communication particularly advantageous in managing big data and enabling real-time communication. The security of quantum communication is grounded in physical laws, ensuring a more secure and reliable method of transmitting information.

With quantum encoding and encryption technologies, quantum communication can prevent unauthorized access and cracking attempts. Due to the intrinsic properties of quantum states—such as their unpredictability and the impossibility of cloning—quantum communication provides absolute security. This makes quantum communication highly promising for applications in critical fields such as finance, healthcare, government, and the military, where secure data transmission is essential.

(2) **The role of single-photon imaging in quantum communication**
A. **Single-photon source serves as the information carrier in quantum communication**

The single-photon source plays a crucial role in quantum communication [22, 23] due to its unique property of emitting only one photon per emission. This ensures that information transmission is based on a single photon, with each photon carrying data in the form of QBs. The key characteristics of single-photon sources—such as their ability to maintain specific phase and polarization properties—help preserve the integrity of the information being transmitted. Because these photons are individual, they are highly resistant to interference from third parties. Any attempt to intercept or tamper with the information will disturb the quantum state of the photon, alerting the receiver to the presence of eavesdropping. This inherent sensitivity to disturbances from external parties guarantees the security of quantum communication, making it highly secure compared to classical communication systems. Thus, the use of single-photon sources in quantum communication ensures that any unauthorized access or manipulation of the transmitted data can be detected immediately.

B. **Single-photon detectors are used to receive and detect quantum signals**

A single-photon detector is a crucial component of a quantum communication system, enabling efficient detection and reception of photon signals for decoding. During the QKD process, the receiver uses the detector to detect the single photon sent by the sender and measure its state to extract key information. The high sensitivity and low noise of single-photon detectors ensure the security of key distribution. If an eavesdropper attempts to intercept the communication, they must measure the photons, which alters the photon's state, allowing the receiver to detect the tampering and immediately be alerted to the eavesdropper's actions.

C. **Single-photon imaging technology realizes tasks such as QKD and quantum teleportation**

QKD [24–26] is a critical technology for ensuring communication security in quantum communication, with the combination of single-photon source and SPD technology being essential for its implementation. By transmitting and detecting single photons, key information can be securely sent and verified. Quantum teleportation leverages quantum entanglement to transmit quantum states to remote locations without transferring the physical carrier. In quantum teleportation, a single-photon source produces two entangled photons, enabling the transmission of quantum information between them. Single-photon imaging facilitates this process, ensuring that remote communication remains highly secure and reliable.

(3) **Technical challenges and solutions in quantum communication**

A. **Improve the purity of the single-photon source**

The single-photon source is a critical component for generating QBs. However, current single-photon sources often face challenges such as low brightness and purity, which limit the transmission distance and communication rate. To address these issues, research is focused on exploring novel material systems like 2D materials and QDs to achieve high-brightness and high-purity single-photon emission. Additionally, optimized single-photon source structures, such as micro-nano structures and photonic crystals, are being designed to enhance brightness and purity. Post-selection techniques using mult-iphoton sources are also employed to select qualified single-photons, further improving the purity of the single-photon source.

B. **Optimize the quantum communication protocol to improve communication efficiency and security**

Quantum communication protocols form the backbone of quantum communication systems. However, existing protocols often face challenges such as low communication efficiency and insufficient security, limiting their practicality and broader application. To address these issues, new protocols like the BB84 and E91 protocols have been developed to enhance both communication efficiency and security. These protocols leverage quantum mechanical properties, such as entanglement and superposition, to enable unconditionally secure key distribution and communication. Additionally, ongoing improvements to existing protocols aim to optimize performance, such as incorporating entanglement exchange technology and quantum repeaters to facilitate long-distance communication and large-scale quantum networks. Security analysis is also essential for identifying potential vulnerabilities and attack methods, leading to the development of countermeasures like stronger authentication and encryption mechanisms to ensure the security of quantum communication.

3. **Future trends of single-photon imaging technology**

(1) **Technological innovation**

A. **Development of a novel single-photon source and detector**

The advancement of single-photon imaging technology is closely linked to innovations in single-photon sources and detectors. Future development will focus on improving efficiency, reducing noise, and increasing the number of probes. Research on new single-photon sources will target materials like QDs, quantum wells, and nanostructures to enhance luminescence efficiency, reduce spectral width, and improve photon emission purity and quantity. Researchers will also explore quantum entanglement and interference for more efficient photon emission. In terms of detectors, future efforts will aim to improve detection efficiency and reduce noise, with a focus on semiconductor nanostructures and superconducting materials. Additionally, novel detection mechanisms, such as single-photon counting and time-resolved detection, will be explored to enhance detector performance.

B. **Design optimizations**

High-performance imaging systems are essential to single-photon imaging technology, and their design and optimization are key to enhancing imaging quality and efficiency. Future advancements will focus on several aspects: First, system design will aim to improve imaging resolution and sensitivity by incorporating novel optical elements and materials, such as super lenses and photonic crystals. Second, efforts will focus on enhancing imaging speed and real-time performance through high-speed optical sensors and parallel processing techniques. Additionally, the system will prioritize stability and reliability by integrating precision control systems and advanced optical isolation measures to reduce external interference and noise.

C. **Advanced image processing and analysis algorithm**

Advanced image processing and analysis algorithms are vital to single-photon imaging technology, and their development is essential for improving imaging quality and efficiency. Future advancements will focus on several areas: First, algorithms will aim to enhance image quality and clarity by incorporating denoising and enhancement techniques to reduce noise and blur. Second, efforts will focus on improving resolution and detail by using image reconstruction and super-resolution algorithms to achieve higher image quality. Additionally, algorithm development will focus on enhancing automatic image processing and analysis capabilities. Machine learning and deep learning algorithms will enable automated classification, recognition, and analysis, boosting the level of automation and intelligence in image processing.

(2) **Futuristic applications**

A. **Interdisciplinary integration and innovative application**

The development of single-photon imaging technology has had a profound impact across many disciplines, with an increasing trend toward interdisciplinary integration. In physics, its combination with quantum science opens new research avenues in quantum imaging, communication, and computing. For example, single-photon imaging enables high-precision quantum state measurements, enhancing both the security of quantum communication and the accuracy of quantum computing. In biology, it allows for high-resolution imaging within living cells, providing deeper insights into life processes. When combined with neuroscience, it aids in studying brain information-processing mechanisms, fostering brain-like computing. In materials science, single-photon imaging helps analyze material microstructures, offering crucial insights for designing new materials. Additionally, it has promising applications in environmental and earth sciences, such as monitoring surface water, air pollution, and assessing the impacts of climate change.

B. **To promote single-photon imaging techniques in a various field**

As the technology matures and costs decrease, single-photon imaging is expected to find widespread applications in various fields. In security monitoring, it can enhance efficiency and security in checks. In healthcare, it can aid in the early diagnosis of diseases, improving diagnostic and treatment outcomes. In transportation, single-photon imaging can enable intelligent traffic control, enhancing road safety and efficiency. Additionally, in culture and art, it can be used for restoring and preserving valuable cultural relics, improving their quality and value. In education, it can make abstract concepts more intuitive through teaching demonstrations, facilitating better understanding.

C. **Implementation in complex quantum information and communication technologies**

Single-photon imaging technology holds significant potential in the field of quantum information and communication. In quantum communication, it can enable high-speed and secure QKD, advancing the practical implementation of quantum communication technology. In quantum computing, it allows for precise measurement and manipulation of QBs, enhancing quantum computing performance. As the technology develops, more complex quantum information processing tasks, such as quantum teleportation and the transmission of quantum entanglement, are expected to be achievable. These advancements will greatly accelerate the progress of quantum information science, providing more efficient and secure methods for information processing and communication.

(3) **Challenges and opportunities**

A. **Overcome technical bottlenecks and reduce costs**

Despite its great potential, single-photon imaging technology still faces several technical challenges in practical applications. First, existing single-photon imaging devices are costly to manufacture and operate, limiting their scalability. To address this, new materials and design solutions are needed to reduce both equipment and operational costs. Second, single-photon imaging is relatively slow, which poses a challenge in applications requiring real-time imaging. Enhancing imaging speed while maintaining high quality is a key area for improvement. Additionally, environmental factors like light scattering and reflection significantly affect imaging accuracy and clarity. Reducing such interference and improving imaging stability are crucial technical challenges to overcome.

B. **Strengthen international cooperation and technology exchanges**

The development of single-photon imaging technology requires significant scientific investment and technological innovation, which is challenging for any single country or region to achieve independently. Therefore, strengthening international cooperation and knowledge exchange is essential for advancing this technology. Collaborative efforts can provide access to additional research funding, advanced experimental equipment, and shared research facilities, improving research efficiency. Moreover, international cooperation fosters exchanges between research teams, stimulates new ideas and creativity, and drives technological innovation.

C. **Seize the historical opportunity of the development of quantum technology**

As a key branch of quantum technology, the development of single-photon imaging technology will benefit from the broader advancements in quantum science. Currently, quantum technology is progressing rapidly, with countries increasing their research investments to compete for technological leadership. China, for example, has achieved significant milestones, such as launching the world's first quantum science experiment satellite, "Micius [27, 28]" and building the longest land-based quantum communication network. The development of quantum technology can address many challenges in single-photon imaging. For instance, quantum entanglement transmission can enhance imaging quality and transmission distance, while quantum computing and simulation technologies can offer more efficient data processing methods. In conclusion, while the future of single-photon imaging faces technical and cost-related challenges, it also presents opportunities through international cooperation and the growth of quantum technology. Overcoming these challenges and seizing these opportunities will drive the rapid advancement of single-photon imaging technology, benefiting humanity in numerous ways.

# References


1. Hadfield RH et al (2023) Single-photon detection for long-range imaging and sensing. Optica 10(9):1124. https://doi.org/10.1364/optica.488853
2. Tiffenberg J et al (2017) Single-electron and single-photon sensitivity with a silicon skipper CCD. Phys Rev Lett 119(13):1–6. https://doi.org/10.1103/PhysRevLett.119.131802
3. Larson AM (2011) Multiphoton microscopy. Nat Photonics 5(1):1. https://doi.org/10.1038/nphoton.an.2010.2
4. Xu C, Nedergaard M, Fowell DJ, Friedl P, Ji N (2024) Multiphoton fluorescence microscopy for in vivo imaging. Cell 187(17):4458–4487. https://doi.org/10.1016/j.cell.2024.07.036
5. Review T (2010) Single-photon generation and detection. https://doi.org/10.1088/0957-0233/21/1/012002
6. Pyle KE, Wu Y-H, M'Closkey RT (2023) A high-precision continuous scan and step scan system for compact spectrometer applications. IEEE/ASME Trans Mechatron 28(6):3329–3338. https://doi.org/10.1109/TMECH.2023.3265643
7. Bille JF (2014) The limits of human vision. In: Frontiers in optics 2014, p FW2F.1. https://doi.org/10.1364/FIO.2014.FW2F.1
8. Fossum ER, Hondongwa DB (2014) A review of the pinned photodiode for CCD and CMOS image sensors. IEEE J Electron Devices Soc 2(3):33–43. https://doi.org/10.1109/JEDS.2014.2306412
9. Pool PJ, Morris DG, Burt DJ, Bell RT, Holland AD, Smith DR (2005) Application of electron multiplying CCD technology in space instrumentation. Focal Pl. Arrays Sp. Telesc. II 5902(January):59020A. https://doi.org/10.1117/12.621627
10. Tulloch SM, Dhillon VS (2011) On the use of electron-multiplying CCDs for astronomical spectroscopy. Mon Not R Astron Soc 411(1):211–225. https://doi.org/10.1111/j.1365-2966.2010.17675.x
11. Moroni GF, Andersson K, Botti A, Estrada J, Rodrigues D, Tiffenberg J (2021) Charge-collection efficiency in back-illuminated charge-coupled devices. Phys Rev Appl 15(6):1. https://doi.org/10.1103/PhysRevApplied.15.064026
12. Zhao Q, Carrascoso F, Gant P, Wang T, Frisenda R, Castellanos-Gomez A (2020) A system to test 2D optoelectronic devices in high vacuum. J Phys Mater 3(3):36001. https://doi.org/10.1088/2515-7639/ab8781
13. Shafana NJ, SenthilSelvi A (2021) Analysis of AI based brain tumor detection and diagnosis. In: 2021 4th international conference on computing and communications technologies (ICCCT), pp 627–633. https://doi.org/10.1109/ICCCT53315.2021.9711914
14. Tobin R, Halimi A, McCarthy A, Laurenzis M, Christnacher F, Buller GS (2019) Three-dimensional single-photon imaging through obscurants. Opt Express 27(4):4590. https://doi.org/10.1364/oe.27.004590
15. Xu F, Wang YQ, Zhang XF, Wang CY (2018) Subpixel three-dimensional laser imaging with a downscaled avalanche photodiode array using code division multiple access. Commun Phys 2(1):1–9. https://doi.org/10.1038/s42005-018-0096-2
16. Tachella J et al (2019) Real-time 3D reconstruction from single-photon lidar data using plug-and-play point cloud denoisers. Nat Commun 10(1):1–6. https://doi.org/10.1038/s41467-019-12943-7
17. Jiang Y, Karpf S, Jalali B (2020) Time-stretch LiDAR as a spectrally scanned time-of-flight ranging camera. Nat Photonics 14(1):14–18. https://doi.org/10.1038/s41566-019-0548-6
18. Sheehan M, Harrison A, Newman P (2014) Automatic self-calibration of a full field-of-view 3D n-laser scanner BT—experimental robotics: the 12th international symposium on experimental robotics. In: Khatib O, Kumar V, Sukhatme G (eds). Springer Berlin Heidelberg, Berlin, Heidelberg, pp 165–178
19. Jiang P-Y, Li Z-P, Ye W-L, Qiu Z, Cui D-J, Xu F (2024) High-resolution 3D imaging through dense camouflage nets using single-photon LiDAR. Adv Imaging 1(1):011003. https://doi.org/10.3788/ai.2024.10001

20. Tan C et al (2022) Design and demonstration of a novel long-range photon-counting 3D imaging LiDAR with 32 × 32 transceivers. Remote Sens 14(12). https://doi.org/10.3390/rs14122851
21. Tobin R, Halimi A, McCarthy A, Soan PJ, Buller GS (2021) Robust real-time 3D imaging of moving scenes through atmospheric obscurant using single-photon LiDAR. Sci Rep 11(1):1–13. https://doi.org/10.1038/s41598-021-90587-8
22. Couteau C et al (2023) Applications of single photons to quantum communication and computing. Nat Rev Phys 5(6):326–338. https://doi.org/10.1038/s42254-023-00583-2
23. Bian L et al (2023) High-resolution single-photon imaging with physics-informed deep learning. Nat Commun 14(1):5902. https://doi.org/10.1038/s41467-023-41597-9
24. Nurhadi AI, Syambas NR (2018) Quantum key distribution (QKD) protocols: a survey. In: 2018 4th international conference on wireless and telematics (ICWT), pp 1–5. https://doi.org/10.1109/ICWT.2018.8527822
25. Yang J et al (2024) High-rate intercity quantum key distribution with a semiconductor single-photon source. Light Sci Appl 13(1):150. https://doi.org/10.1038/s41377-024-01488-0
26. Zapatero V et al (2023) Advances in device-independent quantum key distribution. npj Quantum Inf 9(1):10. https://doi.org/10.1038/s41534-023-00684-x
27. Lu C-Y, Cao Y, Peng C-Z, Pan J-W (2022) Micius quantum experiments in space. Rev Mod Phys 94(3):35001. https://doi.org/10.1103/RevModPhys.94.035001
28. Yin J et al (2019) Quantum science experiments with Micius satellite. In: Conference on lasers and electro-optics, p JTu3G.4

# Chapter 5
# Spectral Imaging Technology

**Abstract** Spectral imaging integrates spectral analysis with digital imaging to capture spatial and spectral information, offering enhanced object analysis beyond traditional methods. It spans wavelengths from ultraviolet to infrared, utilizing technologies like light sources, detectors, and data algorithms. Key systems include grating-based, point-scanning, multispectral, and hyperspectral imaging, each with unique advantages in biomedicine, environmental monitoring, and material analysis. The chapter addresses the transformative role of spectral imaging in medical diagnostics, where it improves tumor detection, disease monitoring, and drug development via non-invasive, real-time monitoring. AI integration further boosts its application in traditional Chinese medicine and food quality control. Spectral imaging promises to revolutionize scientific research, medical diagnostics, and industrial applications in future.

**Keywords** Spectral imaging · Hyper spectral imaging · Multi-wavelength imaging · Machine vision · Biological tissue imaging · Food imaging

## 5.1 Basic Concepts of Spectral Imaging

Spectral imaging [1, 2] is an advanced technology that combines spectral technology and digital imaging technology, enabling it to obtain both spatial and spectral information of objects simultaneously. The following is a detailed explanation of the basic concepts, generation principles, classification, and basic principles and applications of spectral analysis in spectral imaging:

1. **Spectrum**

(1) **The concept of spectrum**

Spectrum refers to the complex color light through the dispersion system (such as prism, grating) separation, the dispersion of the wavelength (or frequency) scattered pattern. In simple terms, the spectrum is the distribution of light at different



S. Du and W. Yin, *Photodetection and Image Sensing Techniques*, Advanced Topics in Science and Technology in China 2, https://doi.org/10.1007/978-981-96-5479-6_5

wavelengths in space. The existence of the spectrum indicates that the light is fluctuating, and the different spectra represent different material composition and structural information.

(2) **Generation mechanism and classification of spectra**

The generation mechanism of spectra is based on light-matter interactions. When light hits matter, it interacts with atoms, molecules, or ions in matter, causing the energy of photons to be absorbed, emitted, or scattered. These processes produce specific spectral signals, the analysis of which allows insight into the nature of matter. Spectra can be classified in different ways. Common classification methods include: (1) emission spectrum: when atoms, molecules, or ions in matter are excited, light of a specific wavelength of light will be emitted to form an emission spectrum. Emission spectra are often used to analyze the elemental composition and excited state properties of matter. (2) Absorption spectrum: When light passes through matter, matter will absorb light at a specific wavelength to form an absorption spectrum. Absorption spectroscopy is often used to analyze the molecular structure and chemical bonding properties of substances. (3) Scattering spectrum: when the light shines on the surface of the matter, the scattering phenomenon will occur to form the scattering spectrum. Scattering spectra are often used to analyze the surface properties and particle sizes of matter, etc.

The basic principle of spectral analysis is to obtain information about the composition, structure and properties of matter by measuring the spectral signals of matter at different wavelengths. Spectral analysis has the advantages of high sensitivity, good selectivity and fast analysis speed, and is widely used in chemistry, physics, biology, medicine and other fields. In spectral imaging, spectral analysis is mainly used to obtain the spectral feature information of objects, so as to achieve high resolution and high sensitivity imaging of objects. By measuring the reflection, emission and transmission spectrum of the object at different wavelengths, the spectral characteristic information of the object can be obtained, and further understand the chemical and physical properties of the object.

2. **Imaging**

(1) **The concept of imaging**

Imaging, which is the process of forming an image, usually refers to the conversion of optical information (such as brightness, color, shape, etc.) into an observable or treatable image by a certain technology or device. In spectral imaging, imaging is not only to obtain the spatial information of the object, but also includes obtaining the spectral information of the object, that is, the spectral characteristics of the reflected, transmitted, or emitted light of the object at different wavelengths.

(2) **Principles of traditional imaging techniques**

Traditional imaging techniques, such as photography, are based on geometric optics. In short, the light forms an image of an object on a sensor (such as a film or image sensor). This image records spatial information about the brightness, color, and shape of the object, but does not contain spectral information.

3. **Spectral imaging**

(1) **The concept of spectral imaging**

Spectral imaging is a technique that combines spectroscopy and digital imaging techniques. It records the reflection brightness distribution or fluorescence brightness distribution of multiple narrow-band monochromatic light in a certain spectral range through a video imaging spectrometer, forming a spectral image set composed of many monochromatic light images.

(2) **The basic mechanism of spectral imaging**

The basic principle of spectral imaging is the scattering and refraction of light. It uses the imaging spectrometer to image the object and obtain the image information of the object in multiple spectral bands simultaneously. The imaging spectrometer is composed of a liquid crystal adjustable wavelength filter (LCTF), digital CCD camera, lighting source and computer and special software. In the imaging process, the computer controls the liquid crystal adjustable wavelength filter successively through the narrow band monochrome light in the predetermined number of equal spacing wavelength positions within a certain range, so that the reflected light or fluorescence of the material in each band reaches the CCD sensor through the filter, forming a spectral image set.

4. **Differences between spectral imaging and traditional imaging techniques**

Traditional imaging technology is mainly based on the RGB (red, green, blue) three-base color model, which captures the color and intensity of the light reflected or through the object, providing a high spatial resolution image. Spectral imaging technology can record a wide range of spectrum from ultraviolet to infrared, and each band can reveal different properties of the object's surface, so it can obtain richer and more accurate information.

Traditional imaging techniques are widely used in everyday photography, basic remote sensing applications, and visual art in general. Spectral imaging technology has wide application prospects in modern science and technology, production, military reconnaissance, target identification, fine agriculture, environmental monitoring, biomedical imaging and many other fields.

Although conventional imaging techniques can provide images with high spatial resolution, they are relatively limited in the capture of spectral information, being limited to the visible spectral range. Spectral imaging technology has both spatial resolution and spectral resolution capabilities, which can provide more comprehensive and more in-depth object information. Spectral imaging provides us with a completely new way of observing and analyzing objects by a combination of spectral and digital imaging techniques with broad applications and potential.

### 5. Composition and classification of spectral imaging

#### (1) Components

##### A. Light source

The light source is the primary component of the spectral imaging system, which is responsible for emitting light radiation, providing energy for the entire spectral imaging process. The choice of light source directly affects the performance of the spectral imaging system. Therefore, factors such as the light source stability, spectral purity, luminescence efficiency, service life and other factors should be considered when selecting the light source. Common light sources include laser light source, LED light source, bulb light source, etc.

##### B. Light separation system

The main function of the spectral system is to divide the light radiation emitted by the light source into several monochromatic lights in a certain way so that the subsequent imaging system can process the light of different wavelengths separately.

##### C. Imaging system

The imaging system is responsible for imaging the monochromatic light separated from the spectral system to obtain spectral images of different wavelengths. The imaging system is usually composed of an optical lens, a reflector lens, an optical path adjustment device, etc. Through the combination of these optical elements, the focusing, reflection, and refraction of light can be realized, to obtain a clear spectral image.

##### D. Detection system

The main function of the detection system is to receive the spectral images formed by the imaging system and to convert them into electrical signals for subsequent data processing and analysis. The detection system usually consists of photo detector, signal amplifier, signal processing device, etc. Photo detectors include PDs, CCD, and CMOS, which can convert optical signals; signal amplifiers and signal processing devices are responsible for magnifying and processing the detected electrical signals to improve the SNR of the spectral imaging system.

#### (2) Classification

##### A. Filtered spectral imaging

Filtered spectral imaging technology is through different wavelengths of light waves through a specific material, due to different wavelengths, resulting in different spectral signals. The filter spectral imaging system is mainly composed of light source, sample, filter, detector, etc. After the light emitted by the light source passes through the sample, the light of a specific wavelength is selected through the filter, which is then received by the detector and converted into an electrical signal, and finally, the spectral information is obtained through data processing. The advantage of filtering spectral imaging is the spectral information with a high SNR, while the disadvantage is the slow imaging speed and the need to replace filters of different wavelengths.

B. **Interferometric spectral imaging**

Interferometric spectral imaging [3, 4] technology uses the principle of interference to analyze the interference effect of optical waves and obtain the spectral information of the sample. The interferometric spectral imaging system is mainly composed of light source, sample, optical path latency, beam splitter, interferometer, detector, etc. After the light from the light source passes through the sample, it is divided into two beams, one through the interferometer and the other one directly to the detector. Two lights meet at the detector to produce the interference phenomenon and the spectral information of the sample. The advantage of interferometric spectral imaging technology is that it can obtain multi-wavelength spectral information in a single exposure, but the disadvantage is that the system is more demanding and the data processing is more complicated.

C. **Dispersive spectral imaging**

Dispersive spectral imaging [5, 6] uses the dispersion characteristics of optical waves in the medium, whose propagation velocity is dependent on the wavelength, to separate light at different wavelengths to obtain spectral information. The dispersive spectral imaging system mainly consists of the light source, sample, dispersion element (such as prism or diffraction grating), detector, etc. After the light emitted by the light source passes through the sample, the light of different wavelengths is separated through the dispersion element, and then received by the detector and converted into the electrical signal, and finally, the spectral information is obtained through data processing. The advantage of dispersion spectral imaging technology is that wide-band spectral information can be obtained in a single exposure, while the disadvantage is that the dispersion elements are highly demanding and susceptible to the influence of dispersion factors such as nonlinearity.

D. **Computational spectral imaging**

Computational spectral imaging is to process the acquired spectral information mathematically to obtain the spectral information of the sample. The computational spectral imaging system mainly consists of light source, sample, detector, data processing algorithm, etc. After the light emitted from the light source passes through the sample, it is received by the detector and converted into electrical signals, and then the spectral information is extracted through the data processing algorithm. The advantages of computational spectral imaging are the flexible selection of the bands of interest, the disadvantage of the strong dependence on data processing algorithms and the relatively low requirements for hardware equipment.

6. **Technical characteristics of spectral imaging**

(1) **High-spectral resolution**

High-spectral resolution is an important feature of spectral imaging techniques. It refers to the ability of spectral imaging devices to distinguish very close spectral wavelengths, usually to nanometer levels. This high-resolution feature gives spectral imaging extremely high accuracy when analyzing the spectral characteristics of matter. Due to the small differences in the spectral characteristics of different substances, the high spectral resolution can help us to distinguish these differences, and thus achieve the accurate identification of substances. This has wide applications in chemical composition analysis, environmental monitoring, biomedicine and other fields. For example, in environmental monitoring, the atmosphere can be accurately resolved by high spectral resolution; in the biomedical field, hyperspectral imaging technology [7, 8] can be used for the diagnosis and analysis of cancer tissue.

(2) **Spatial resolution**

Spatial resolution is another important feature of spectral imaging techniques. It refers to the ability of a spectral imaging device to distinguish two adjacent targets, usually expressed by angle or distance. Higher spatial resolution means that spectral imaging devices can resolve smaller targets, allowing for finer spatial information. The high spatial resolution enables spectral imaging in target detection and mapping. For example, in the field of remote sensing, spectral imaging technology with high spatial resolution can help us obtain detailed information about the earth's surface, including topography, landforms, vegetation cover, etc., which is of great significance for geological exploration, urban planning, agricultural monitoring, etc.

(3) **Rich information content**

Spectral imaging technology has rich information content because of its ability to obtain both image information and spectral information of the target. By analyzing this information, we can obtain detailed information about the type, content and distribution of the target material, which is of high value for research and application in many fields. In chemistry, spectral imaging can be used to analyze the composition and structure of substances; in biomedicine, it can be used to diagnose diseases and monitor treatment effects; in environmental monitoring, it can be used to monitor the distribution and spread of pollutants. These applications all benefit from the rich information content of spectral imaging techniques.

(4) **Real-time**

Real-time performance is yet another important feature of the spectral imaging technology. It refers to the ability of spectral imaging devices to obtain spectral information about the target in a short time, usually at the millisecond or second level. This real-time nature makes spectral imaging a significant advantage in scenarios that require rapid response. For example, in the field of safety inspection, spectral imaging can be used to quickly detect dangerous goods; in the medical field, it can be used to monitor patient physiological parameters in real-time; in the industrial field, it can be used to monitor the production process in real-time. These applications require real-time spectral imaging to obtain information and process it.

## 5.2 Classification and Principle of the Spectral Imaging System

1. **The basic principles of the spectral imaging system**

The spectral imaging system combines spectral analysis and imaging techniques to obtain both spatial and spectral information of objects simultaneously. Here is a detailed explanation of the basic principles of the spectral imaging system:

(1) **Principles of spectral analysis**

Spectral analysis is a method to study the composition and structure of light by measuring its absorption, emission, or scattering. In spectral imaging, the principle of spectral analysis is mainly reflected in the measurement and analysis of the spectral properties of objects at different wavelengths. When light passes through the matter, the matter absorbs light at a specific wavelength, forming an absorption spectrum. By analyzing the absorption spectra, the composition and concentration of the substance can be determined. When the matter is excited, it emits light at a specific wavelength, forming an emission spectrum. By analyzing the emission spectrum, the composition and structure of the material can be determined. When the incident light interacts with the matter, the scattering phenomenon occurs. We can determine the composition and structure of matter by measuring the scattering angle and the intensity of the scattered light. In spectral imaging, these spectral analysis methods are used to measure and analyze the reflection, transmission, or emission spectra of an object at different wavelengths to obtain information about the spectral features of the object.

(2) **Imaging mechanism**

The imaging mechanism refers to the process of transforming the information of an object into visual images or reflected information through the principles of optics or electronics. In spectral imaging, the imaging principle is mainly reflected in the imaging of the object through the imaging equipment (such as the imaging spectrometer), and the spatial information and spectral information of the object are presented in the form of the image. Imaging equipment usually includes lenses, filters, detectors, and other components. The lens is used to focus the light reflected or emitted by an object, the filter is used to select the light at a specific wavelength, and the detector is used to convert the optical signal into an electrical signal to form a digital image. By adjusting the wavelength of the filter, the spectral image of the object at different wavelengths can be obtained to realize the spectral imaging.

### (3) Mechanism of spectral imaging

The mechanism of spectral imaging refers to the combination of spectral analysis and imaging technology to achieve high spectral-spatial resolution imaging of objects. In spectral imaging systems, spectral analysis and imaging techniques are interdependent and mutually reinforcing. On the one hand, spectral analysis provides rich spectral information for imaging, making the imaging results contain not only the spatial information of the object but also its spectral features. This information can be used to identify the material, color, structure and other properties of the object, providing strong support for scientific research and technical applications. On the other hand, the imaging technology provides the spatial resolution for the spectral analysis. By dividing the object into pixel units, the spectral information of each pixel unit can be measured separately to achieve high-spectral spatial resolution imaging of the object. This imaging modality can reflect the spectral properties of the object more finely and improve the accuracy and reliability of spectral analysis.

The basic principle of the spectral imaging system is to combine spectral analysis and imaging techniques to achieve high spectral-spatial resolution imaging of objects. This imaging modality not only has a wide application prospect but also provides powerful tools and support for scientific research and technology applications.

## 2. Classification of the spectral imaging systems

### (1) Classification based on the spectral elements

#### A. Prism spectral imaging system

The prism spectral imaging [9] system mainly divides the incident light through the prism and realizes the spectral separation of light according to the different refraction angles of different wavelengths in the prism. Such systems usually have the advantages of simple structure and low cost but have relatively low resolution and efficiency. In prism spectral imaging systems, multi-piece prism combinations are usually used to improve the effect of spectral separation. Such systems are widely used in astronomical observation, biomedical imaging and other fields.

#### B. Grating spectral imaging system

The grating spectral imaging [10] system divides the incident light through a grating, which has highly regular transparent and opaque stripes and can separate the incident light according to the wavelength. Such systems have high resolution and efficiency, but they have a relatively complex structure and high cost. In grating spectral imaging systems, diffraction gratings or shining gratings are often used to achieve efficient spectral separation. This system is widely used in optical instruments, laser technology, optical fiber communication and other fields.

### (2) Classification based on the detection mode

#### A. Point-scanning spectral imaging system

A point-scanning spectral imaging system is an imaging system that obtains spectral information by scanning the sample area point by point. In such systems, the detector of the spectrometer scans the sample point by point while recording the spectral information at each point. This spectral information is then converted into images, enabling spectral imaging of the sample. The advantage of the point-scanning spectral imaging system is its high spectral and spatial resolution. Because the system can obtain the spectral information of the sample point by point, high-quality spectral images can be obtained. Moreover, such a system can also adjust the spatial resolution of the spectral image by varying parameters such as the scanning speed and step size. However, the disadvantages of point-scanning spectroscopic imaging systems are the relatively slow scanning speed and require a long data acquisition time. This makes the system limited in practical applications.

#### B. Face array spectral imaging system

The array spectral imaging system uses the array detector to obtain the spectral information of the sample simultaneously. Such a system enables rapid imaging by distributing the spectral information of the sample over a 2D surface array. Compared with point scanning spectral imaging systems have faster data acquisition speed and higher efficiency. The advantages of the surface array spectral imaging system lie in its fast imaging capability and high efficiency. Because the system can obtain the spectral information of the entire sample area simultaneously, the data acquisition time is greatly shortened. This makes the surface array spectral imaging system more extensive in practical applications. However, the disadvantages of face array spectral imaging systems are their relatively low spectral resolution and spatial resolution. Because the system distributes the spectral information of the sample over a 2D plane array, there are some limitations in the spectral resolution and spatial resolution.

In general, based on the classification of detection methods, spectral imaging systems are mainly slightly scanning and surface array. The point scanning spectral imaging system has high spectral resolution, but slow scanning speed; the surface array spectral imaging capability and high efficiency, but relatively low spectral resolution and spatial resolution. According to the practical application requirements, an appropriate spectral imaging system can be selected.

### (3) Classification based on spectral resolution

#### A. Multispectral imaging system

The multispectral imaging system is a spectral imaging system that obtains the spectral information of the target object in different bands. Such systems usually include detectors in multiple bands, such as visible light, near-infrared, mid-infrared, etc. The main feature of the multispectral imaging system is the ability to obtain the radiation information of the target object in different spectral bands, to realize the spectral analysis of the target object. Multispectral imaging systems are widely used, such as monitoring crop growth and pest detection, environmental monitoring, disease diagnosis and drug development.

B. **Hyperspectral imaging system**

The hyperspectral imaging system is a spectral imaging system that can obtain the spectral information of the target object in the entire spectral range. It typically includes detectors of hundreds or even thousands of bands, capable of obtaining full spectral information of the target object from ultraviolet to far-infrared. The main feature of the hyperspectral imaging system is the ability to obtain more detailed spectral information of the target object, thus achieving accurate identification and analysis of the target object. Because the hyperspectral imaging system can obtain information about the target object in the entire spectral range, it can be used in many fields, such as geological exploration, biomedicine, environmental monitoring, etc. For example, in geological exploration, hyperspectral imaging systems can be used to identify different mineral components to help geologists better understand geological structures; in biomedical exploration, hyperspectral imaging systems can be used to non-invasively diagnose diseases, such as by analyzing spectral information from tissue to identify cancer cells.

(4) **Other classification methods**

A. **Real-time spectral imaging system and non-real-time spectral imaging system**

Real-time spectral imaging system refers to the spectral imaging system for real-time monitoring of the samples during the data acquisition process. Such systems are usually used in situations where dynamic change processes are monitored, such as chemical reaction process monitoring, biomedical imaging, etc. The real-time spectral imaging system can respond quickly and display the spectral information of the sample in real-time, but it may be complicated in data processing and analysis. The non-real-time spectral imaging system is a system that analyzes the spectral information of the sample offline after obtaining all the data. This system is suitable for high accuracy and stability of spectral data, such as scientific research, material analysis, etc. The non-real-time spectral imaging system is relatively simple in data processing, but cannot monitor the dynamic changes of samples in real time.

B. **Visible light spectral imaging system and infrared spectral imaging system**

Visible light spectral imaging system refers to a spectral imaging system with a working band in the range of visible light (about 400–760 nm). Such systems are usually applied to sample analysis with obvious color changes, such as plant leaf spectral analysis, art identification, etc. The visible light spectral imaging system has a relatively low equipment cost, but it is greatly disturbed by strong light sources such as sunlight. The infrared spectroscopic imaging system works in the infrared band (about 760 nm to 1 mm), and is suitable for the analysis of non-transparent samples, such as rock, soil, biological tissue, etc. The infrared spectroscopy imaging system is sensitive to the thermal radiation of samples, so it is widely used in remote sensing and geological exploration. However, due to the weak light source in the infrared band, the system has higher technical requirements for noise suppression and signal enhancement.

C. **Static spectral imaging system and dynamic spectral imaging system**

A static spectral imaging system refers to a system in which a sample is aged once within a certain period. This system is suitable for analyzing stable and unchanged samples, such as material analysis and composition detection in the laboratory. Simple static spectral imaging system data acquisition but cannot capture the sample over time. Dynamic spectral imaging systems can image the sample multiple times in continuous time to observe the change of the sample over time. Such systems are suitable for the analysis of dynamically changing samples, such as chemical reaction processes, metabolic processes within organisms, etc. Dynamic spectral imaging systems can provide complete dynamic information of samples, but the data acquisition and processing are more complex, which requires high stability and accuracy of the system.

3. **The key technology of the spectral imaging system**

(1) **Design and manufacture of high performance optical spectral components**

The high-performance spectroscopic element, the core component of the spectral imaging system, is responsible for separating the optical signal by wavelength for subsequent spectral analysis and imaging. The design and manufacture of high-performance spectroscopic elements need to consider multiple factors, including the selection of materials, the structural design of the components, the optimization of optical properties, etc. Common spectral elements include filters, grating, prism, etc., in which filters can realize the selection of specific bands while grating and prism can achieve more refined spectral resolution.

(2) **Research and development of high-sensitivity and high-resolution detectors**

The detector is another important component of the spectral imaging system that is responsible for converting optical signals into electrical signals for digital processing and storage. The high-sensitivity and high-resolution detector can capture the optical signals more accurately and improve the performance of the spectral imaging system. Developing a detector with high sensitivity and high resolution needs to consider the material, structure, manufacturing process and other factors to improve the response speed, SNR and resolution.

(3) **Efficient data acquisition and processing algorithm**

Spectral imaging systems produce large and complex data volumes, thus requiring efficient data acquisition and processing algorithms to process and analyze these data. Data acquisition algorithms need to ensure the accuracy and integrity of the data while minimizing data redundancy and noise. The processing algorithm requires data pre-processing, feature extraction, classification and identification, to extract useful information. These algorithms need to be designed and optimized for the characteristics of spectral imaging systems to improve the efficiency and accuracy of data processing.

(4) **Stable imaging system and control system**

The imaging systems and control systems are the framework of the spectral imaging systems, which are responsible for maintaining the stable operation of the system and ensuring the accuracy and reliability of the imaging results. The imaging system needs to ensure the stability of the lens, the accuracy of the optical path, and the response speed of the detector to obtain high-quality image data. The control system needs to accurately control each part of the system, including the filter switch, detector calibration, etc., to ensure the stability and reliability of the system.

The key technologies of the spectral imaging system cover the design and manufacturing of optical separation components, the research and development of detectors, data acquisition and processing algorithms, and the imaging system and control system. The continuous development and optimization of these technologies will further improve the performance and range of applications of spectral imaging systems.

## 5.3 Data Processing and Analysis of Spectral Imaging

1. **Characteristics of the spectral imaging data**

(1) **Data dimension**

Spectral imaging data usually have a 3D data structure, including two spatial dimensions and one spectral dimension [11]. Two spatial dimensions correspond to the rows and columns of the image, namely the spatial resolution of the image, and the spectral dimension corresponds to the spectral information of each pixel at different wavelengths, namely the spectral resolution. This 3D data structure enables the spectral imaging data to contain both rich spatial and spectral information.

(2) **Data volume**

Because spectral imaging data has a 3D data structure, and each pixel needs to record spectral information at multiple wavelengths, the amount of data is usually large. As the spectral resolution improves and the imaging range expands, the data volume also increases exponentially. This huge amount of data put forward higher requirements for data storage, transmission and processing.

(3) **Data quality**

The quality of spectral imaging data is affected by various factors, including the performance of the imaging system, environmental conditions, properties of the target object, etc. Among them, the performance of the imaging system is one of the key factors, including the optical performance of the lens, the detector sensitivity, the spectral resolution, etc. At the same time, environmental conditions can also affect the imaging quality, such as light conditions, atmospheric absorption, and scattering. In addition, the properties of the target object can also affect the imaging quality, such as the reflectivity, transmission, fluorescence properties, etc. Therefore, when obtaining spectral imaging data, these factors need to be considered comprehensively and measures taken accordingly to improve data quality.

To improve the quality of the spectral imaging data, the preprocessing of the raw data is usually required, such as noise removal, distortion of correction, registration, etc. Meanwhile, regular calibration and maintenance of the imaging system are also required to ensure the stability and reliability of its performance. In addition, some advanced algorithms and techniques can be used to post-process and analyze the data to extract more useful information.

2. **Preprocessing of the spectral imaging data**

(1) **Data correction**

Data correction is an important step in the preprocessing of spectral imaging data, and its main purpose is to eliminate or reduce errors and uncertainties in the data and improve the quality and reliability of the data. Data correction mainly includes the following aspects:

A. **Band correction**

Band correction addresses the band interference in multispectral or hyperspectral imaging data. The spectral properties may vary in different bands, resulting in data errors during transmission, detection and recording. The purpose of band correction is to have the same spectral characteristics, thus improving data comparability and analytical accuracy. Band correction methods include linear transformation, log transformation, power transformation, etc.

B. **Radiation correction**

Radiation correction is the process of converting imaging data from radiance values to reflectivity or absorbance. Due to natural factors such as the atmosphere and clouds, the radiation component included in the imaging data may deviate from the actual surface reflectance or absorption rate. The purpose of the radiation correction is to remove this bias and make the data more truly reflect the spectral properties of the surface.

C. **Geometry correction**

Geometric correction is the coordinate transformation of imaging data to eliminate or reduce geometric distortions caused by the curvature of the earth, sensor attitude changes, topographic fluctuations and other factors. The purpose of geometric correction is to make the data with a unified coordinate system for subsequent data analysis and processing. Geometric correction methods include projection transformation, affine transformation, and spherical interpolation.

D. **Atmospheric correction**

Atmospheric correction is corrected for the effect of the atmosphere on spectral imaging data. Aerosols, water vapor and other components in the atmosphere may absorb or scatter light, resulting in a decrease in data quality. The purpose of atmospheric correction is to eliminate or reduce this effect and improve the authenticity of the surface reflectivity or absorption rate of the data. Atmospheric correction methods include dark field correction, atmospheric correction model (such as the 6S model), etc.

(2) **Data calibration**

Data calibration is another critical step in the preprocessing of spectral imaging data, whose primary purpose is to convert the data into absolute or relative standard units to facilitate subsequent data analysis and application. Data calibration mainly includes the following aspects:

A. **Absolute calibration**

Absolute calibration is the process of converting spectral imaging data into absolute radiation measurements. This step usually requires calibration using standard boldface or other known spectral radiation sources. By measuring the spectral radiation intensity of these standard sources, the correspondence between the spectral radiation intensity and the image value can be established to achieve the absolute calibration of the data. The main purpose of absolute calibration is to eliminate the nonlinear errors and system noise of the sensor response and improve the accuracy and reliability of the data.

B. **Relative calibration**

Relative calibration refers to the inter-band scale adjustment of spectral imaging data to eliminate radiation intensity nonlinear errors and system noise between different bands. This step is usually performed by minimizing the sum of residual squares between the bands. The main purpose of relative calibration is to improve the interpretability of the data and ensure that the proportion of radiation intensity between different bands is consistent during subsequent analysis and application.

C. **Band balancing**

Band balancing is a linear combination of bands between spectral imaging data to eliminate non-linear errors in radiation intensity and system noise between bands. This step is usually performed by minimizing the sum of residual squares between the bands. The main purpose of band matching is to improve the accuracy and reliability of the data and to ensure that the radiation intensity between the different bands is consistent during the subsequent analysis and application.

D. **Scale calibration**

Scale calibration is the radiometric scale or data format required to convert spectral imaging data into a specific application. This step usually requires the design according to the practical application requirements and the sensor characteristics. The primary purpose of scaling calibration is to ensure that the data meets the needs of specific applications during subsequent analysis and application, such as converting the data into reflectance, transmission, or other physical measurements.

In conclusion, data calibration is a key step in the preprocessing process of spectral imaging data. Through absolute calibration, relative calibration, band leveling and scale calibration, the accuracy and reliability of data can be effectively improved, to ensure that the data has high value in the subsequent analysis and application process.

(2) **Data enhancement**

Data enhancement is the last step in the preprocessing of spectral imaging data, and its main purpose is to improve the usability and reliability of the data through a series of processing methods. Data enhancement mainly includes the following aspects:

A. **Noise suppression**

Noise is an important factor affecting the quality of spectral imaging data, so noise suppression is essential during data enhancement. Common noise suppression methods include wavelet transform, median filter, Gaussian filtering and so on. These methods can improve the quality of the data by smoothing the data and removing random noise.

B. **Detection and removal of outlier values**

In the spectral imaging data, the presence of outliers can affect the analysis and processing results of the data. Therefore, outliers need to be detected and removed during data enhancement. Common outlier detection methods include the Z-score method, the boxplot method, etc. Once outliers are detected, they can be removed by interpolation, deletion, etc.

C. **Data interpolation**

In practice, there may be missing values in the spectral imaging data for various reasons. To make full use of these data, it needs to be filled in completely by the interpolation method. Common interpolation methods include linear interpolation, spline interpolation, Kriging interpolation, etc. The interpolated data can reflect the actual spectral information more accurately, thus improving the reliability of the data.

D. **Feature extraction**

Feature extraction is the process of extracting information from raw data useful for analysis and processing. In spectral imaging data enhancement, feature extraction can help us discover potential rules and relationships in the data, and thus provide strong support for subsequent analysis. Common feature extraction methods include principal component analysis (PCA), linear discriminant analysis (LDA), etc. Through feature extraction, we can reduce the dimension of the data, simplify the data model, and improve the analysis efficiency.

Data enhancement is an important link in the pre-processing of spectral imaging data. Through a series of processing methods such as noise suppression, outlier detection and removal, data interpolation and feature extraction, the quality and availability of data can be effectively improved and provide a reliable basis for subsequent analysis and processing.

3. **The feature extraction of the spectral imaging data**

(1) **Spectral characteristics**

Spectral feature extraction is an important part of spectral imaging data processing, and its main purpose is to extract useful information from complex spectral data for subsequent analysis and application. Spectral feature extraction mainly relies on spectral analysis algorithms, including spectral curve fitting, spectral classification, spectral angle normalization, and principal component analysis. Spectral curve fitting is to fitting the spectral curve through a mathematical model to extract the key parameters in the curve. These parameters can reflect the physical and chemical properties of the matter, such as the wavelength, strength, etc. Spectral classification is performed by comparing the spectral curves of different samples, and dividing them into different categories. This method is often used in areas such as species identification and environmental monitoring. Spectral angle normalization is a method to transform spectral data into angle data, and the differences between them can be extracted by comparing the spectral angles of different samples. PCA is a dimensionality reduction method that extracts the most important features by projecting the raw spectral data into a new coordinate system.

(2) **Spatial characteristics**

Spatial feature extraction refers to extracting information related to spatial location from spectral imaging data. This information can reflect the spatial distribution and morphological characteristics of the features. The methods of spatial feature extraction can mainly include image segmentation, edge detection, morphological processing, etc. Image segmentation divides the image into areas, each representing an object or object. This method can be used to extract specific regions in the image for subsequent processing and analysis. Edge detection is used to detect the boundaries of the objects in an image. Through edge detection, the contour and shape information of objects can be extracted, which can be used for object recognition and classification. Morphological processing is an image processing method based on mathematical morphology, which can extract the specific shape and structure information in the image through expansion, corrosion and open operation.

(3) **Characteristics of the spectral-space combination**

Spectral-space joint feature extraction is the simultaneous extraction of spectral and spatial features from spectral imaging data. This method can more comprehensively reflect the properties and distribution, thus improving the accuracy of feature identification and classification. The methods of spectral-spatial joint feature extraction mainly include spectral-spatial correlation analysis and spectral-spatial clustering analysis. Spectral-spatial correlation analysis is to extract the correlation between spectral and spatial features. This method can be used to find the association rules between spectral and spatial features, and thus for the identification and classification of ground features. Spectral-spatial cluster analysis combines spectral and spatial features to conduct a cluster analysis of samples. This method can be used to find samples with similar spectral and spatial features, and thus for the identification and classification of ground features.

4. **Analysis method of spectral imaging data**

(1) **Cluster analysis**

Cluster analysis is one of the important methods of spectral imaging data analysis, mainly by dividing the samples in the dataset into different categories according to their similarity. In the analysis of spectral imaging data, cluster analysis can help us discover potential rules in the data, remove redundant information, reduce the dimensionality of the data, and facilitate subsequent analysis and processing. Common clustering analysis algorithms include the K-means algorithm, hierarchical clustering algorithm, DBSCAN algorithm and so on. Each of these algorithms has its advantages and disadvantages, and they need to be selected according to the specific situation in practical application. For example, the K-means algorithm is easy to implement and easy, but susceptible to the initial center point; the hierarchical clustering algorithm can automatically determine the number of clusters, but the calculation is high in complexity; the DBSCAN algorithm can identify the cluster of any shape, but the parameter adjustment is complicated. In practical applications, cluster analysis usually needs to be used in combination with other methods, such as feature selection, dimension reduction, etc., to improve the effect of the analysis.

(2) **Classification and identification**

Classification identification is another important task for spectral imaging data analysis, mainly by building models to classify unknown samples. In the analysis of spectral imaging data, classification identification can help us identify different substances and analyze the composition and structure of substances. Common classification and recognition algorithms are support vector machines (SVM), decision trees, random forests, neural networks and so on. Each of these algorithms has its advantages and disadvantages, and they need to be selected according to the specific situation in practical application. For example, the SVM algorithm has good performance when handling small samples and high-dimensional data, requiring multiple parameters; the decision tree algorithm is simple to interpret but easy to overfit; the random forest algorithm can effectively avoid overfitting but has high computational complexity; neural network algorithm has good generalization ability, but requires a lot of training data and tuning parameters. In practical applications, classification recognition usually needs to be used in combination with other methods, such as feature selection and feature extraction, to improve the effect of recognition.

(3) **Spectral unmixing**

Spectral unmixing is an important task in the analysis of spectral imaging data, mainly in the decomposition of the mixed spectra into multiple original spectra. In practical applications, spectral unmixing can help us to analyze the different components in the mixture, thus realizing the qualitative and quantitative analysis of the mixture. Common spectral unmixing algorithms include non-negative matrix decomposition (NMF), independent component analysis (ICA), linear discriminant analysis (LDA) and so on. Each of these algorithms has its advantages and disadvantages, and they need to be selected according to the specific situation in practical application. For example, the NMF algorithm can effectively deal with nonlinear problems, but has high computational complexity; the ICA algorithm can effectively separate mixed signals, but is vulnerable to noise; the LDA algorithm can effectively improve the classification effect, but requires a lot of training data. In practice, spectral unmixing usually needs to be used in combination with other methods, such as feature selection, dimension reduction, etc., to improve the effect of unmixing.

(4) **Spectral reconstruction**

Spectral reconstruction is an important task in the analysis of spectral imaging data, mainly through the known partial spectral information, to reconstruct the complete spectral information. In practice, spectral reconstruction can help us recover spectra contaminated by noise and improve the quality of spectral data. Common spectral reconstruction algorithms include the least squares method, maximum likelihood method, sparse representation and so on. Each of these algorithms has its advantages and disadvantages, and they need to be selected according to the specific situation in practical application. For example, the least squares algorithm is simple to calculate but susceptible to noise; the maximum likelihood algorithm can effectively improve reconstruction but has high computational complexity; the sparse representation algorithm can effectively handle high-dimensional data but requires a lot of training data. In practical applications, spectral reconstruction usually needs to be used in combination with other methods, such as feature selection, dimension reduction, etc., to improve the effect of reconstruction.

## 5.4 Application of Spectral Imaging in Remote Sensing, Biomedicine, and Other Fields

1. **Application of spectral imaging in the field of remote sensing**

(1) **Vegetation monitoring**

Spectral imaging technology is an imaging technology that can obtain the spectral information of the target object, which has wide applications in the field of remote sensing. In terms of vegetation monitoring, spectral imaging technology can obtain information on the species, health status and growth status of vegetation by analyzing the spectral characteristics of vegetation. Different types of vegetation have different spectral properties, so they can be classified by spectral imaging techniques. For example, the healthy vegetation performs differently from the spectral images. By analyzing the spectral images, it can be determined whether the vegetation is infected by pests and diseases. Vegetation coverage refers to the proportion of vegetation coverage on the surface, which is an important indicator to measure the growth status of vegetation. Spectral imaging technology can calculate the cover of vegetation by monitoring the spectral reflectance of vegetation. In addition, by analyzing the spectral curve of vegetation, we can understand the growth status of vegetation, such as chlorophyll content, and vegetation density. The vegetation information obtained by spectral imaging technology can provide important data support for environmental protection and resource management. For example, spectral imaging technology plays an important role in forest fire early warning, wetland protection, and grassland resource surveys. Spectral imaging technology is widely used in vegetation monitoring, which provides us with detailed, accurate and real-time vegetation information, which is of great significance for ecological environment protection and the rational utilization of resources.

(2) **Water body monitoring**

Water body monitoring is one of the important applications of spectral imaging technology in the field of remote sensing. Through spectral imaging technology, we can monitor and analyze the color, transparency, and pollutant content of water in detail. First, the spectral imaging technology can help us to analyze the color of water bodies. Under different water quality conditions, the reflection and absorption characteristics of the water body spectrum are different, therefore, through spectral imaging technology, we can get the color information of the water body, to judge the quality of water quality. For example, clean water bodies often appear blue or green, while heavily polluted water bodies may appear brown or black. Secondly, spectral imaging techniques can be used to monitor the transparency of the water bodies. The transparency of the water body is related to the reflection and scattering characteristics of the light spectrum. By analyzing the spectral data, we can get the transparency information of the water body. The higher the transparency, the clearer the water, and the more turbid the versa. In addition, spectral imaging techniques can also be used to monitor the pollutant content in water bodies. The reflection and absorption characteristics of the spectra are different among different pollutants, so by analyzing the spectral data, we can get information on the type and content of pollutants in the water. This is of great significance for the protection and management of water resources. Finally, spectral imaging techniques can also be used to monitor the degree of eutrophication in water bodies. Eutrophication refers to the amount of nutrients in water, which leads to the deterioration of water quality. The higher the degree of eutrophication in water, the worse the water quality. Through spectral imaging technology, we can obtain information on the degree of eutrophication of water bodies, thus providing important data support for the conservation and utilization of water resources. Spectral imaging technology has wide application prospects in water monitoring, which can help us better understand the quality of water bodies and provide a scientific basis for the protection and treatment of water resources.

(3) **Geological exploration**

Spectral imaging technology plays an increasingly important role in the field of geological exploration. By analyzing the spectral characteristics of the ground, this technology can reveal multi-dimensional information such as the type, structure and composition of the ground, and greatly promotes the depth and accuracy of geological exploration. Spectral imaging enables comprehensive recording and analysis of the reflection, transmission and emission spectra of objects. By processing and parsing these spectral data, scientists can identify unique spectral features of different features, such as mineral composition, soil type, and vegetation cover, etc. For example, in remote sensing images, spectral imaging technology can distinguish minerals such as opivine, pyroxene, and feldspar, which is an important tool for geologists to identify and analyze geological structures. Through spectral imaging technology, different kinds of ground features can be identified and distinguished, including different types of rock, soil, water bodies, etc. At the same time, spectral imaging can also reveal the internal structural information of the ground object, such as cracks, pores, etc. This information is essential for assessing the engineering nature and resource potential of geological bodies. Spectral imaging technology can not only provide static geological information but also dynamically monitor the changes of ground objects. For example, through long time series spectral imaging data, erosion and deposition can be monitored, which is of great significance for land resource management, environmental protection and geological disaster prevention. Spectral imaging technology provides accurate data support in mineral resources exploration, reservoir evaluation, environmental monitoring and other aspects. It helps geological explorers to predict more accurately and evaluate the distribution of resources, and improve the efficiency and benefit of resource development. At the same time, through monitoring environmental changes, spectral imaging technology also provides a scientific basis for environmental protection and sustainable development. In conclusion, the application of spectral imaging technology in geological exploration has greatly improved the accuracy and efficiency of geological exploration and has an irreplaceable role in the rational development of resources and the effective protection of the environment.

(4) **Urban planning and management**

In the field of urban planning and management, spectral imaging technology is playing an increasingly important role. This technology can provide urban planners with rich information on land use, greening status, building distribution and so on by conducting detailed spectral monitoring and analysis of different areas of the city. First, in terms of land use, spectral imaging technology can help planners to accurately judge the use status of urban land, including different functional areas such as residential, commercial and industrial areas. Analyzing the spectral characteristics of different regions, can reveal the changing trend of land use and provide a scientific basis for the rational allocation and planning of land resources. Secondly, spectral imaging technology is equally important for the monitoring of urban greening conditions. It can assess the urban greening coverage rate and monitor the plant species and growth status, to provide decision support for urban greening planning and construction. For example, by analyzing the spectral data of green space, the layout of green space can be optimized and the quality of urban greening can be improved. In addition, the spectral imaging technology also enables a detailed analysis of urban buildings. It can identify building materials, structures and health conditions and provide important information for the maintenance and management of urban buildings. At the same time, through the integration and analysis of the building spectral data, the layout and design of urban buildings can be optimized, and the utilization efficiency of urban space can be improved. Spectral imaging has wide applications in urban planning and management. It can not only provide detailed data support for urban planners but also help them to better understand and respond to the various challenges in urban development. By making full use of spectral imaging technology, we can build more habitable, efficient and sustainable cities.

2. **Application of spectral imaging in the biomedical field**

(1) **Medical imaging diagnosis**

The application of spectral imaging technology in the field of medical imaging diagnosis is increasingly expanding, and its technical characteristics and advantages have revolutionized doctors and patients. Compared with traditional medical imaging technologies such as X-ray, CT scanning and MRI, these technologies usually can only provide a single dimension of information. Spectral imaging technology can realize multi-parameter and multi-scale imaging of biological tissues by obtaining spectral information of different wavelengths. This ability allows the spectral imaging technology to show great potential and value in tumor diagnosis, cardiovascular disease monitoring, and the detection of neurological diseases. In tumor diagnosis, spectral imaging technology enables the qualitative and quantitative analysis of tumors by analyzing the spectral properties of tumor tissue. Because of the differences in optical properties of tumor versus normal tissue, spectral imaging techniques can take advantage of these differences to identify and locate tumor tissue. By detecting specific biomolecules and metabolites in tumor tissue, spectral imaging can not only help doctors determine the presence of the tumor but also provide the biological properties of the tumor, such as malignancy, aggressiveness and whether metastasis occurs. This information is crucial for doctors to plan treatment and determine prognosis. Moreover, the application of spectral imaging techniques in cardiovascular diseases is equally important. Cardiac and vascular diseases often involve complex biomolecular and cellular processes, and spectral imaging enables the monitoring of these changes non-invasively. For example, monitoring of blood oxygenation, cholesterol, and inflammatory markers by spectral imaging techniques can help in the early detection of heart disease. During surgery, spectral imaging can also help doctors understand the state of the tissue in real-time, reducing the risk of surgery. Spectral imaging similarly shows its unique advantages in the detection of neurological diseases. For example, in the diagnosis of brain tumors, stroke, and Alzheimer's disease, spectral imaging can provide more accurate information on brain tissue metabolism and cytostructure than traditional imaging techniques. This helps neurologists to judge the condition more accurately and develop personalized treatment plans. The application of spectral imaging technology in medical imaging diagnosis has greatly improved the accuracy and efficiency of medical diagnosis by providing more accurate and comprehensive information. With the continuous development and improvement of this technology, it will be more widely used in the biomedical field in the future and is expected to make a greater contribution to the cause of human health.

(2) **Drug development and evaluation**

With the rapid development of science and technology, the gradual improvement of the medical level, and the continuous improvement of people's health expectations, the requirements for the accuracy, speed, sensitivity and intelligence of medical testing are also becoming higher and higher. Although conventional traditional imaging examinations such as physical and chemical examination, ultrasound, X-ray, electronic computed tomography (CT) and MRI have good diagnostic reference value, they will bring a certain degree of damage to the patient's body, such as ionized electromagnetic radiation, which may bring secondary trauma to the patient. Although positron emission computed tomography (PET-CT) has the characteristics of specificity, high safety and whole-body imaging, it is difficult to be accepted by ordinary patient groups because of its high examination cost. In drug analysis, gravimetric analysis, acid–base titration, pH measurement, electrophoresis, chromatography, and mass detection, although the accuracy is good, but also has certain limitations in different degrees, such as mass spectrometry in gas chromatography-mass spectrum (GC–MS) method analysis to sugar, benzene compounds weak response, mobile phase can only be limited to volatile components. The samples analyzed by liquid chromatography-mass spectrometry (LC–MS) should preferably be aqueous solution or methanol solution, and the LC mobile phase should not contain nonvolatile salt. Even the use of mass spectrometry detectors, later maintenance, and operation specifications have high requirements, and it is difficult to roll out in a large range at this stage. Hyperspectral Imaging technology, as an emerging means of analysis, has made rich research results in geological surveys, soil and vegetation ecological monitoring, Marine detection, food safety, archaeology and cultural museums in recent years. HSI technology [5, 12, 13] is the rich spectral information and image information combined from 2D spatial image data combined with one-dimensional spectral data and 3D information, with timing, positioning, simplicity, intelligence, and accuracy, is widely used in the early modern medical disease screening and diagnosis, clinical guidance treatment, clinical efficacy evaluation, sample collection information time less than 10 s, almost can do the "real-time" application. In the aspect of traditional medicine, observation, listening, inquiry and diagnosis, as a key step in TCM diagnosis, is to observe the god, color, shape, state and tongue image of the patient, and to judge the lesions of zang-fu organs. Years of medical experience make the diagnosis more accurate, which is difficult to describe and quantify, and has been criticized by Western medicine. The emergence of HSI technology, combined with artificial intelligence and deep learning technology, transforms the subjective observation of TCM into a quantifiable data processing, matching and diagnosis process, which greatly improves the accuracy and credibility of observation. In addition, HSI technology has shown a blowout development in many fields such as forensic identification, histopathological analysis, quality control, source identification and so on (Table 5.1 and Fig. 5.1).

(3) **Biological tissue imaging**

Biological tissue imaging [14–16] refers to the use of spectral imaging technology to obtain the spectral characteristics of biological tissues, to achieve the nondestructive detection and imaging of biological tissues. In the field of biomedicine, spectral imaging technology is widely used, spectral imaging technology can obtain the spectral characteristics of biological tissues and, through the analysis of spectral characteristics, can achieve the early diagnosis of disease. For example, spectral imaging technology can be used to detect skin cancer, breast cancer and other diseases, and realize the early detection and early treatment of diseases. Spectroscopic imaging technology allows for real-time monitoring of drug distribution and delivery processes in living organisms. Through spectral imaging technology, real-time monitoring of drug delivery can be achieved to optimize the drug delivery protocol and improve the therapeutic efficacy of drugs. Spectral imaging technology can obtain the spectral properties of biological tissues, and through the analysis of the composition of biological tissues can be realized. For example, using spectral imaging technology can analyze water, protein, fat and other components in biological tissues, and provide important information for clinical diagnosis and treatment. Spectroscopic imaging techniques can be used to analyze gene expression in biological tissues. Through spectral imaging technology, the quantitative analysis of gene expression in biological tissues can be realized, providing important information for studying the function of biological tissues and the occurrence mechanism of diseases. Spectroscopic imaging techniques can enable non-destructive imaging of biological tissue. Through spectral imaging technology, the structural information of biological tissues can be obtained, providing an important reference for clinical diagnosis and treatment. Spectroscopic imaging technology is widely used in the biomedical field and can be used in disease diagnosis, drug delivery, biological tissue component analysis, gene expression analysis and other aspects, providing strong technical support for the realization of non-destructive detection and imaging of biological tissues.

(4) **Physiological and metabolic monitoring**

The application of spectral imaging technology in the field of physiological and metabolic monitoring is increasingly highlighting its unique value. This technology enables non-invasive and damage-to-free detection of biological tissues, providing strong technical support for the early diagnosis and treatment of diseases. In terms of physiological monitoring, spectral imaging technology can monitor physiological parameters such as blood flow, oxygen saturation and nerve conduction in real-time. For example, through spectral imaging technology, the flow of blood in the body and the oxygen saturation change can be observed in real-time, which has important

**Table 5.1** Application of HSI technology in the pharmaceutical field

| Class (category) | Scope of application (scope of application) |
|---|---|
| Clinical Medicine | Vascular disease-peripheral artery and coronary artery disease (vascular disease-peripheral arterial disease and coronary artery disease), arthritis (arthritis), retinal, brain, and kidney diseases (retinal, brain and kidney disease), intraoperative identification of tissue organs and nerve fibers (intraoperative identification of tissues, organs and nerve fibers) |
| | Dental caries (dental caries), laryngeal lesions (laryngeal lesions), ophthalmopathopathy-retinopathy (ophthalmic diseases-retinopathy), skin lesions-burns (skin injury-thermalinjury |
| | Diabetic complications-diabetic foot (diabetic disease-diabitic foot), intestinal ischemia (intestinal ischemia), haemorrhagic shock (hemorrhagic shock), renal transplantation (renal transplantation), liver resection (hepatectomy), hyperlipidemia (hyperlipoidemia) |
| | Brain tumors (brain tumors), brain cancer (brain cancer), lung adenocarcinoma (lung adenocarcinoma), tongue cancer (tongue cancer), breast cancer (breast cancer), gastric cancer (gastric cancer), colorectal cancer (colorectal cancer), cervical tumors (cervical tumors), cervical cancer (cervical cancer), melanoma (melanoma), skin cancer (skin cancer), prostate cancer (prostate cancer), ovarian cancer (ovarian cancer), lymphid carcinoma (lymph cancer) et al. |
| Traditional Chinese Medicine | Table tification (superficies syndrome), in card (interior syndrome), cold syndrome (cold syndrome), thermal certificate (heat syndrome), false evidence (deficiency syndrome), empirical evidence (sthenic syndrome), negative syndrome (yin syndrome), positive certificate (yang syndrome), organ diseases (viscera disease) [such as heart disease (heart disease), pulmonary fever (lung heat disease), stroke (stroke), etc.] acupuncture and physical therapy (acupuncture-therapy) |
| Basic Medicine—Forensic Medicine and Histopathology | Biological sample identification (identification of biological samples), blood (blood), saliva (saliva), semen (sperm), urine (urine), histopathological section analysis (histopathological section analysis) cell diagnosis of microscopic lesions (diagnosis of microcellular lesions) |
| Drugs and their toxic and harmful residues | Prevention (Radix Stephaniae Tetrandrae), fennel (Foeniculum vulgare Mill.), Pon (Radix Ophiopogonis), Bitter almond (Semen Armeniacae Amarum), Peach kernel (peach seed), Honeysuckle (Flos Lonicerae), Ginseng Ginseng (Ginseng Radix et Rhizoma), Cordyceps sinensis (Cordyceps sinensis), Loquat leaves (Folium Eriobotryae), Ginkgo biloba leaves (Ginkgo biloba), Mulberries (mulberry), Black barbary wolfberry (Lycium ruthenicum), Preparation of Chinese patent medicines (preparation of proprietary Chinese medicine), pesticide and heavy metal residues in medicinal materials (residues of pesticides and heavy metals in medicinal materials) |

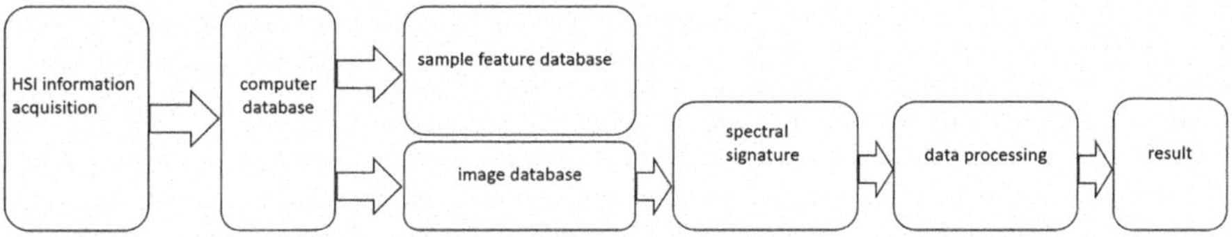


**Fig. 5.1** Flow chart of HSI technical data analysis

significance for understanding the state of blood circulation in the human body, as well as the discovery and diagnosis of cardiovascular diseases. In terms of metabolic monitoring, spectral imaging enables real-time monitoring of metabolites in living organisms to understand the metabolic state of the organism. For example, by spectral imaging technology, the blood glucose level in the organism can be monitored in real-time, and the trend of blood glucose has important guiding significance for the diagnosis and treatment of diabetes. In addition, spectral imaging technology can also be used to monitor the distribution and efficacy of drugs in living organisms, which is of great reference value for drug development and application. Conclusively, the application of spectral imaging technology in physiological and metabolic monitoring provides strong technical support for the early diagnosis and treatment of diseases, as well as an important reference for the development and application of drugs.

(5) **Food testing**

As a common spectral technology, near-infrared light is an electromagnetic wave that can be used for material quantitative analysis and structural processing and presents different propagation characteristics in different structural materials. To detect food, we can make a qualitative and quantitative analysis of the sample structure and characteristics according to the characteristic function relationship, distinguish different components according to the change of properties, and generate a spectral model. By comparison with known spectra, model correction can be completed to obtain food detection data. With good conductivity, the near-infrared spectrum can be used to carry out multi-channel detection in the process of food production, strengthen the food quality control of each node, and not bring harm to the material structure of the food. Without pretreatment, it can improve the testing efficiency. At the same time, the equipment maintenance cost and energy consumption are low, which will not affect the surrounding environment. It belongs to the green testing technology and meets the requirements of green food production and processing.

In the field of food testing, this technology is mostly used for alcohol and brewing food detection. For example, the detection of ethanol and phenol content in alcoholic food can scientifically predict the best brewing time. In the detection of beer products, the content of reduced sugar, ethanol and other substances can be determined to confirm whether it can meet the food production standards. The composition of edible vinegar is relatively complex, so the traditional detection technology needs to pretreat the sample to complete the complex detection steps, but the application of near-infrared spectroscopy technology can quickly complete the determination of reduced sugar, total ammonia and other components. In addition, the technology can also be used in the detection of dairy products and meat products. For example, to detect cattle dairy products, the identification model can be used to evaluate the nutrient content in the food, and determine the reduced milk content, to complete the identification of food adulteration. The detection of beef and other meat products can obtain substance content data through spectral analysis, which is used to determine food tenderness, fatty acids and other conditions, and complete the detection of excessive drug residues to ensure food quality. However, near-infrared spectroscopy is prone to the environment and measurement conditions of the sample, and the spectral peak overlap may occur in the detection of multiple components. Because in the detection process, a gene may be absorbed by multiple wavelengths, resulting the measurement accuracy is closely related to the nature of the sample, and ultimately causes the accuracy of the detection results is affected. Therefore, for food detection, more spectral models of known samples should be added to effectively improve the accuracy of detection results.

## References

1. Lemke M, Quinlan D (1992) Virtual shared grids based programming environment for architecture-independent development of structured grid applications. In: Lecture Notes in Computer Science (including Subseries Lecture Notes Artificial Intelligence and Lecture Notes in Bioinformatics), vol 634, LNCS, no 1, pp 121–126. https://doi.org/10.1007/3-540-55895-0_406
2. Yang J, Cui K, Huang Y, Zhang W, Feng X, Liu F (2023) Deep-learning based on-chip rapid spectral imaging with high spatial resolution. Chip 2(2):100045. https://doi.org/10.1016/j.chip.2023.100045
3. Itoh K, Inoue T, Ichioka Y (1988) Interferometric spectral imaging in the visible and near-infrared regions. In: Space optics for astrophysics and earth and planetary remote sensing, p WB6. https://doi.org/10.1364/SOA.1988.WB6
4. Yoshimori K (2001) Interferometric spectral imaging for three-dimensional objects illuminated by a natural light source. J Opt Soc Am A 18(4):765–770. https://doi.org/10.1364/JOSAA.18.000765
5. Kumar A, Saxena S, Shrivastava S, Bharti V, Kumar U, Dhama K (2016) Hyperspectral imaging (hsi): applications in animal and dairy sector. J Exp Biol Agric Sci 4(4):448–461. https://doi.org/10.18006/2016.4(4).448.461
6. Chen Z, Wang Q, Zhang H, Nie P (2021) Hyperspectral imaging (HSI) technology for the non-destructive freshness assessment of pearl gentian grouper under different storage conditions. Sensors 21(2). https://doi.org/10.3390/s21020583
7. Yoon J (2022) Hyperspectral imaging for clinical applications. BioChip J 16(1):1–12. https://doi.org/10.1007/s13206-021-00041-0
8. Xue Q et al (2021) Underwater hyperspectral imaging system using a prism–grating–prism structure. Appl Opt 60(4):894–900. https://doi.org/10.1364/AO.415351
9. Dirk CW, Delgado MF, Olguin M, Druzik J (2009) A prism–grating–prism spectral imaging approach. Stud Conserv 54(2):77–89. https://doi.org/10.1179/sic.2009.54.2.77
10. Zhang X, Li X, Tang X (2022) Study on a plane grating spectral imaging system with smile and keystone eliminated. J Opt Technol 89(5):262–268. https://doi.org/10.1364/JOT.89.000262

11. Darma PN, Baidillah MR, Sifuna MW, Takei M (2020) Real-time dynamic imaging method for flexible boundary sensor in wearable electrical impedance tomography. IEEE Sens J 20(16):9469–9479. https://doi.org/10.1109/JSEN.2020.2987534
12. Biswas A (2024) Real-time dynamic imaging : revolutionizing medical visualization, vol 16, pp 188–189. https://doi.org/10.47532/1755-5191.2024.16(3).188-189
13. Liu H et al (2024) Learning-based real-time imaging through dynamic scattering media. Light Sci Appl 13(1):194. https://doi.org/10.1038/s41377-024-01569-0
14. Gustin Nurdialit D, Mujtahid Anas A, Mitrayana M (2023) Photoacoustic microscopy system for biological tissue imaging. J Phys Conf Ser 2498(1):12016. https://doi.org/10.1088/1742-6596/2498/1/012016
15. Susaki EA et al (2020) Versatile whole-organ/body staining and imaging based on electrolyte-gel properties of biological tissues. Nat Commun 11(1):1982. https://doi.org/10.1038/s41467-020-15906-5
16. Wu D et al (2021) Imaging biological tissue with high-throughput single-pixel compressive holography. Nat Commun 12(1):4712. https://doi.org/10.1038/s41467-021-24990-0

# Chapter 6
# 2D Imaging Technology

**Abstract** This chapter highlights the role of 2D imaging in machine vision, safety inspection, and security, emphasizing the importance of dynamic range, resolution, and deep learning in enhancing the performance of machine vision systems. High-resolution 2D imaging enables accurate detection, object classification, and real-time monitoring, making it essential for multiple industries. Dynamic range extension enhances image contrast and brightness, improving system reliability under varying lighting conditions. In safety inspection systems, 2D imaging, integrated with multi-modal fusion technologies like X-ray, infrared, and millimeter-wave imaging, offers enhanced detection capabilities. AI-driven analysis, big data, and automated systems significantly reduce human error, accelerate detection, and improve efficiency in security checks. These advancements contribute to airport security and industrial automation.

**Keywords** 2D imaging · Acoustic imaging · Holographic imaging · Multimodal fusion imaging · Magnetic resonance imaging · Infrared imaging · X-ray imaging · Microscopic imaging

## 6.1 The Principles 2Dimensional Imaging

1. **Basics of one and 2D imaging**

(1) **Principles of optical imaging**

The principle of optical imaging [1–5] involves light propagation, lens imaging and camera imaging. Here are some detailed explanations of these three aspects:

A. **Light transmission**

Light is an electromagnetic wave, traveling at a speed of about 300,000 km/s in the air. In different media, such as air, water or glass, the speed of the propagation of light changes, depending on the refractive index of the medium. When light moves from one medium to another, its propagation direction changes, a phenomenon called the refraction of light. The law of refraction describes the behavior of light at the

 
S. Du and W. Yin, *Photodetection and Image Sensing Techniques*, Advanced Topics in Science and Technology in China 2, https://doi.org/10.1007/978-981-96-5479-6_6

interface of two different media, namely, the incidence angle, the refraction angle, and the refractive index of the two media. The light is also reflected when it encounters the surface of the medium. According to the law of reflection, the incident light reflected light and normal are located in the same plane, and the incident angle is equal to the reflection Angle.

B. **Lensing imaging**

A lens is an optical element used to control the propagation of light. It can be a convex lens (thick center, thin edge) or a concave lens (thin center, thick edge) [6]. The imaging principle of a convex lens is that when the light passes through the lens, the light is refracted due to the difference in the media on both sides of the lens. For a convex lens, the refracted light converges toward the central axis of the lens, forming a real or virtual image. The real image can be projected on the screen, while the virtual image can only be seen directly by the observer. The imaging principle of a concave lens is opposite to that of a convex lens. It refracts light through refraction, forming a virtual image. Concave lenses are often used to correct vision problems such as myopia.

C. **Camera imaging**

A camera is a device that uses optical imaging principles for image capture. It is mainly composed of a lens, shutter, sensor and image processing system. The lens is the main optics of the camera, which receives the light and focuses it on the sensor. Parameters such as focal length, aperture and focal adjustment can affect the quality and effect of the imaging. The photoreceptor is a key component of the camera and is responsible for converting optical signals into electrical signals. There are two common sensors: CCD and CMOS. Parameters such as the resolution and sensitivity of the photosensitive component determine the sharpness and quality of the camera image. The shutter controls the length of camera exposure. During the exposure, the sensor receives the light from the lens and records the image information of the object. The length of the exposure time can affect the brightness and detail performance of the image. The image processing system is responsible for converting the electrical signals captured by the sensor into digital images and performing various image processing and optimization operations, such as noise reduction, sharpening, color correction, etc. The resulting digital image can be stored in the camera's memory card or sent to a computer or other device via a transmission interface for further processing and analysis.

(2) **Principles of electronic imaging**

A. **The electron-beam scan**

Electron beam scanning is one of the core steps of electron imaging technology. During imaging, the electron beam (composed of high-energy electrons) is generated and focused through scanning systems (such as the electron gun and electromagnetic lens). These electron beams are precisely controlled to scan the target object or imaging media in a point-by-point or line-by-line manner. During scanning, the electron beam interacts with the target object, producing secondary electrons, X-rays, or other forms of radiation. These secondary radiations carry information about the surface or internal structure of the target object. Parameters such as the speed, precision and energy of electron beam scanning have important effects on the imaging quality. By adjusting these parameters, image resolution, contrast and depth can be controlled.

B. **Pixel array**

The pixel array is part of an electronic imaging system used to receive and record information about electron beam scans. It is usually composed of a series of closely packed pixel cells, each of which independently detects and quantifies the signals generated by the electron beam. The design of pixel arrays and material selection are critical for imaging quality. The pixel array with high sensitivity, high resolution and low noise can provide clearer and more detailed images. In electronic imaging systems, pixel arrays often work in synchronization with the electron beam scanning system. As the electron beam is scanned, the pixel array sequentially records information about each pixel and eventually forms the complete image.

C. **Image processing**

Image processing is another important link in electronic imaging technology. It involves enhancing, filtering, segmentation and recognition of the raw image data to extract useful information or improve the visual effect of the image. The image processing algorithm can be customized and optimized according to the specific requirements. For example, in medical imaging, special algorithms may be needed to highlight diseased areas or increase the contrast of an image; in industrial detection, algorithms may be required to detect defects of an object or to measure its dimensions. The development of image processing technology has brought more possibilities for electronic imaging technology. Through advanced image processing technology, more valuable information can be extracted from electronic imaging data and applied in various fields.

(3) **Acoustic imaging principle**

A. **Sonic propagation**

A sound wave is a type of mechanical wave that requires a medium to propagate, such as air, water, or solid. Sound waves are generated by object vibration and propagate through the particles in the medium that collide with each other. The velocity, direction, and intensity of the acoustic wave propagation are affected by the properties of the medium. For example, in solids, acoustic waves usually travel faster than in gas. In addition, sound waves may encounter reflection, refraction, diffraction and other phenomena during the propagation process. Acoustic wave propagation is fundamental to acoustic imaging. By measuring the propagation characteristics of the sound waves in the medium, information about the position and characteristics of the sound source or the target object can be obtained.

B. **Ultrasonic imaging**

Ultrasonic imaging uses the propagation and reflection properties of ultrasonic waves in the medium to obtain information about the internal structure of the object. The ultrasonic frequency is higher than the hearing range of the human ear, with better directionality and penetration. During the ultrasound imaging process, the ultrasound emitter fires an ultrasonic pulse to the target object. Reflections occur when ultrasound encounters different media or internal structures of objects. These reflected waves are received by the receiver and converted into electrical signals, which then undergo signal processing and analysis to generate an image of the target object. Ultrasonic imaging is widely used in medical diagnosis and nondestructive testing. For example, in medicine, ultrasound imaging can be used to observe the fetus, examine internal organs, etc.; in nondestructive testing, ultrasound imaging can be used to detect internal defects of the material, evaluate the material properties, etc.

C. **Acoustic holographic imaging**

Acoustic holographic imaging [7, 8] is an advanced acoustic imaging technology, which uses the principle of interference to obtain all the information of the sound field (amplitude distribution and phase distribution) of the observed object. The basic principles of acoustic holographic imaging are similar to those of optical holographic imaging. During the acoustic holographic imaging process, the sound field distribution can be reconstructed by measuring the propagation and reflection of sound waves around the object. Then use the interference principle to stack and interfere the sound signals of different positions, so as to obtain the holographic image of the object. Acoustic holographic imaging has the advantages of high resolution and high sensitivity, which can be applied to nondestructive destructive detection and sound field analysis. For example, in nondestructive detection, acoustic holographic imaging can be used to detect the shape and size of defects inside the material; in acoustic field analysis, acoustic holographic imaging can be used to analyze the distribution and characteristics of the sound field.

2. **The applications of 2D imaging technology**

(1) **Applications of optical imaging technology**

A. **Daily life**

The optical imaging technology is very useful in daily life. The most common are digital cameras and mobile phone cameras, which use optical imaging technology to capture images and allow us to record every moment in our lives. In addition, optical imaging technology is also used in home projectors, TVs, computer monitors and other devices, so that we can enjoy a high-quality visual experience.

B. **Scientific research**

In the field of scientific research, optical imaging technology also plays an important role. Microscope is a typical application of optical imaging technology in the field of scientific research, which can help scientists to observe the microscopic cell and tissue structure, and then further study the various phenomena of biology. In addition, telescopes are also one of the important applications of optical imaging technology, which allows us to observe distant stars and cosmic phenomena, providing an important tool for astronomical research.

C. **Industrial testing**

Optical imaging technology also plays a key role in the field of industrial detection. For example, in the manufacturing process, the optical imaging technology can be non-contact, and high precision detection, such as the application of an optical projector in the manufacturing of liquid crystal display, can ensure that the quality of the product reaches high standards. In addition, optical imaging technology is also widely used in barcode scanning, machine vision detection and other fields, greatly improving industrial production efficiency.

(2) **Applications of electronic imaging technology**

Electronic imaging technology is widely used in medical imaging, semiconductor detection, and machine vision:

A. **Medical imaging**

Electronic imaging technology has a crucial role in the field of medical imaging. With X-rays, CT, MRI and other devices, doctors can obtain high-definition, high-resolution images of the patient's body. These images can not only help doctors accurately diagnose the disease, but also provide precise navigation during the procedure. For example, virtual reality technology can combine surgical simulation with real-world surgical practice to project computer-displayed virtual models onto patient entities, helping doctors to locate and guide the surgery and reduce surgical risk.

B. **Semiconductor detection**

During the semiconductor manufacturing process [9], electronic imaging technology is used to detect defects and impurities in wafers. Devices such as scanning electron microscopy (SEM) can be imaged at high resolution to detect micron-level defects. In addition, electronic imaging technology can also be used to analyze the materials and structures of semiconductors, to provide an important reference for semiconductor design and manufacturing.

C. **Machine vision**

Machine vision [10] is an important application area of electronic imaging technology. In industrial automation and robotics, machine vision systems can capture images through cameras and use image processing algorithms to analyze and identify images. For example, on an automated production line, a machine vision system can detect features such as the size, shape, and color of a product for automatic sorting and recognition. In addition, machine vision can also be used in face recognition, license plate recognition and other security fields. Electronic imaging technology plays an important role in the fields of medical imaging, semiconductor detection and machine vision, providing strong support for the development of modern technology and industry.

(3) **Application of acoustic imaging technology**

A. **Medical diagnosis**

Acoustic imaging technology has important applications in the field of medical diagnosis, especially ultrasonic imaging technology. Ultrasound examination is a non-invasive, radiation-free examination method to observe the structure and function of the internal organs in the human body through the reflection and propagation of sound waves. Doctors can use ultrasound imaging technology to detect the lesions of the fetus, heart, liver, kidney, abdomen and other organs, which has the advantages of no pain, no radiation, and no need for injection of contrast agent. Additionally, advanced techniques such as acoustic force microscopy can also be used to observe the morphological and structural information of cells and tissues, helping doctors in pathological analysis and diagnosis.

B. **Non-destructive testing**

Nondestructive testing refers to the use of the tested object, using the change of abnormal internal structure or defects of heat, sound, light, electricity, or magnetic reaction, to the detection of internal and surface defects, and the type, nature, quantity, shape, position, size, distribution and change of defects. Acoustic imaging technology, especially ultrasound imaging technology, is in an important position in the field of nondestructive testing. Through the propagation and reflection characteristics of ultrasonic waves in the material, the defects, inclusions and cracks inside the material can be detected, providing an important basis for product quality control and safety assessment.

C. **Marine exploration**

Acoustic imaging technology also plays an important role in the ocean exploration field. Sonar technology is one of the commonly used acoustic imaging techniques in ocean exploration, which uses ultrasound to detect underwater objects. We can determine the distance between the object and the sound source by calculating the time difference between the sending and receiving the sound waves twice. In addition, acoustic imaging can help determine deep-sea topography and discover undersea volcanoes, faults, and other geological features. In ocean engineering, acoustic technology can also help build stability models and health management systems for marine facilities. In addition, acoustic communication technology is also one of the important applications in ocean exploration. It can replace electromagnetic wave communication in high-pressure and long-distance transmission scenarios, and provide data transmission, imaging, or control services for deep-sea detection robots and observation stations.

3. **Development trends of 2D imaging technology**

(1) **High-resolution imaging**

High-resolution imaging technology is an important development direction for 2D imaging technology. With the progress of science and technology, new high-resolution imaging technologies are constantly emerging, such as electron microscopy, optical microscopy, and interferometric imaging technologies. These technologies have broad applications in materials science, biology and medicine. The electron microscopy can provides higher resolution and even atomic-level imaging. Among them, the continuous development of scanning electron microscope (SEM) and transmission electron microscope (TEM) provides researchers with a deeper means to observe and understand the micro world. The optical microscope significantly improves the resolution by improving the brightness of the light source and reducing the interference of the stray light. For example, techniques such as laser confocal microscopy and fluorescence microscopy, which can achieve high-resolution imaging of live cells, provide strong support for biological research. Interferometric imaging technology uses the principle of light interference to obtain high-resolution images by comparing the phase differences of two or more light fields. The application of interferometric synthetic aperture radar (InSAR) and interferometric light microscope has made the high-resolution imaging technology achieve remarkable results in earth observation, biomedicine and other fields.

(2) **Real-time dynamic imaging**

Real-time dynamic imaging technology is another important direction of 2D imaging technology. The development of this technology allows researchers to observe dynamic processes in real-time to better understand how things change and develop. High-speed video imaging technology is the key to realizing real-time dynamic imaging. By improving the frame rate of the camera device, real-time imaging of high-speed moving objects can be realized. For example, high-speed cameras that shoot thousands of frames per second can capture the moment when the bullet hits the chamber, or the details of the bird's flight. In addition, real-time dynamic imaging technology also involves the methods of image processing and analysis. By processing and analyzing the continuously taken images, the motion trajectory, speed and acceleration can be extracted, to achieve a deep understanding of the dynamic process.

(3) **Multimodal fusion imaging**

Multimodal fusion imaging [11] is a combination of multiple imaging technologies to make full use of the advantages of various technologies to achieve more comprehensive and accurate imaging. For example, combining optical imaging and MRI technology provides both structural and functional information. This multimodal fusion imaging technology has broad applications in neuroscience, tumor diagnosis and other fields. In addition, the multimodal fusion imaging technology can also achieve imaging at different scales. For example, combining optical imaging and electron microscopy can enable imaging from cell to molecular scale, providing more comprehensive information for biomedical research.

(4) **Intelligent imaging**

Intelligent imaging technology refers to the use of computer technology and artificial intelligence algorithms to realize the automation and intelligence of the imaging process. For example, by exploiting deep learning algorithms, the automatic recognition and classification of images can be realized. This intelligent imaging technology has a wide application prospect in image processing, target detection, and other fields. In addition, intelligent imaging technology can also realize the automatic optimization of the imaging parameters. By exploiting machine learning algorithms, imaging parameters can be automatically adjusted to achieve high-quality imaging. In conclusion, the development trends in 2D imaging technology include high-resolution imaging, real-time dynamic imaging, multimodal fusion imaging, and intelligent imaging. The development and application of these technologies will bring more efficient and accurate imaging solutions in areas such as scientific research and industrial applications.

## 6.2 Classification and Characteristics of 2D Imaging Systems

1. **Classification of one-and 2D imaging systems**

(1) **Optical imaging system**

An optical imaging system is a device that uses optical components to capture, transmit, process, and display images. Here is a detailed explanation of the optical imaging system:

A. **Definition**

Optical imaging systems form images based on the propagation and reflection properties of light. After the light is reflected or transmitted from the observed object, it is focused through optical components such as the lens, and then converted into electrical signals by the photosensitive sensor (such as CCD or CMOS), and performs signal processing and display.

B. **Classification**

(a) **Visible light imaging system**

Visible light imaging systems utilize light in the range of wavelength visible to the human eye. This system is widely used in photography, video surveillance, machine vision and other fields. The characteristic of visible light imaging is that it can capture and restore the color, texture and other visual information of objects, providing people with intuitive and easy-to-understand images.

(b) **Infrared imaging system**

Infrared imaging system [12] uses the infrared radiation emitted by objects for imaging. Because infrared radiation is not restricted by natural conditions such as day and night, sunny days and rainy days, infrared imaging systems can be used at any time and under any weather conditions. Infrared imaging system has the characteristics of high resolution and high sensitivity and can identify and display specific information such as very small objects, lesions, or damage. In addition, infrared imaging technology can also present the distribution and change of thermal radiation, which is also very helpful for some thermal convection, temperature distribution, object heat transfer and other studies.

(c) **Ultraviolet imaging system**

The UV imaging system utilizes UV light for imaging. Such systems are often used for the preventive detection of power equipment, such as corona, flashover, or arc discharge phenomena in substations and high-voltage transmission lines. UV imager has the characteristics of high sensitivity, accurate corona positioning, strong anti-interference ability, can work all day long, and the detection speed is very fast, the test sensitivity is also very high. In addition, due to the use of special filtering technology, the corona discharge detection works to avoid the interference of solar radiation can be detected under solar strong light.

C. **Features**

Optical imaging systems can obtain image information without direct contact with the observed object, which is very useful for some objects or environments that cannot be directly contacted. The optical imaging system can obtain high-resolution images, making the observation and analysis more accurate. Optical imaging systems are suitable for a wide range of imaging requirements from microscopic to macroscopic and can be applied to different fields and scenarios. In conclusion, optical imaging systems have wide application prospects and important roles in modern scientific and industrial fields.

(2) **Electronic imaging system**

A. **Definition**

An electronic imaging system is a system that imagines electronic signals. Based on the change in the electromagnetic properties of materials, electronic imaging can be a physical process or a chemical process. It is widely used in the fields of plate making, proofing, printing, medicine and nondestructive detection.

B. **Classification**

(a) **Scanning electron microscopy (SEM)**

SEM [13] uses the focused high-energy electron beam to scan the sample, through the interaction between the beam and the material, stimulates a variety of physical information, the information is collected, amplified, and imaged, to achieve the purpose of the microscopic morphology of the matter. SEM has the characteristics of high resolution, intuitive imaging, strong 3D sense, and wide magnification range. It has been widely used in life science, physics, chemistry, earth science, materials science and industrial production.

(b) **Transmission electron microscopy (TEM)**

TEM [14] can see subtle structures less than 0.2 μm that cannot be seen under a light microscope, which are called sub-microstructures or ultra-structures. TEM uses the electron beam as the light source, and its wavelength is much shorter than visible and ultraviolet light, so its resolution is much higher than that of optical microscopy. TEM is an important analytical method in many scientific fields related to biology and physics, such as cancer research, virology, materials science, nanotechnology, and semiconductor research.

(3) **The X-ray imaging system**

The X-ray imaging system is a special detection instrument, mainly used in the field of biology. It uses the ability of X-rays to penetrate objects to form images by detecting the difference in their absorption of X-rays. The X-ray imaging system has the characteristics of high definition and remote network diagnosis, which is especially suitable for X-ray imaging in biological research.

A. **Features**

Both SEM, TEM, and X-ray imaging systems provide high-resolution images that facilitate detailed observation and analysis of the structure and properties of the target object. Electronic imaging systems often do not require direct contact with the target object, avoiding possible physical damage and contamination. Electronic imaging systems are suitable for a variety of materials and scenarios, including biological samples, metallic materials, ceramics, and nanomaterials. The electronic imaging system can not only be used for morphology analysis but also for composition analysis, crystal structure analysis, and dynamic process observation.

(4) **Acoustic imaging system**

A. **Definition**

The acoustic imaging system is a system that generates a characteristic image of the internal/external structure of the object by receiving and analyzing the echo or scattering signal of sound waves (such as ultrasonic waves, acoustic waves, etc.) when propagating inside or outside of the object. This technology is widely used in medical diagnosis, non-destructive testing, ocean detection and other fields.

B. **Classification**

(a) **Ultrasonic imaging system**

Using the reflection, scattering and other phenomena generated when the ultrasonic wave propagation inside the object, the image of the internal structure is generated by receiving and analyzing these signals. Ultrasound imaging systems are widely used in the field of medical diagnosis, such as echocardiography, and ultrasound liver examination. It can also be used for industrial nondestructive testing, such as the detection of material's internal defects.

(b) **Sonar system**

Using the propagation characteristics of sound waves in water or other liquid media, the underwater objects are detected by transmitting sound waves and receiving a reflected echo. Sonar system is mainly used in ocean exploration, underwater terrain surveys, underwater communication and other fields.

C. **Features**

The acoustic imaging system can obtain internal or external structural information without direct contact with the measured object, avoiding possible physical damage and contamination. Ultrasonic and sound waves have good penetration and can penetrate many non-metallic materials, such as plastic, rubber, and glass, so acoustic imaging systems are suitable for a variety of materials and scenarios. The acoustic imaging system can obtain and analyze acoustic signals in real-time, to generate image information of objects in real-time and provide support for rapid decision-making. Compared with radiation imaging technologies such as X-ray and CT, acoustic imaging systems have a low safety risk and less impact on the human body and the environment. Acoustic imaging system is a promising technical system. Its non-contact, strong penetration, real-time and high safety characteristics make it have unique advantages in medical diagnosis, nondestructive testing, ocean detection and other fields.

(5) **Other imaging systems**

A. **Magnetic resonance imaging (MRI)**

MRI is a medical imaging technique that uses strong magnetic fields and radiofrequency pulses to generate internal images of the human body. Unlike X-ray and CT scans, ionizing radiation is not used in the MRI. The basic mechanism of MRI is that the resonance frequency of hydrogen atoms in the external magnetic field matches the frequency of the radio frequency pulse to produce a signal. After processing, these signals can generate images of cross-sectional, sagittal and coronal sections of the human body. MRI has a very high soft tissue resolution ability, which can clearly show the structure of the brain, spinal cord, joints, muscles and other tissues. Moreover, MRI can also perform functional imaging, such as fMRI, to observe the brain blood flow changes when performing different tasks to study brain function. MRI is limited by its slow speed and may not be practical enough for emergency cases.

B. **Radioactive imaging**

Radioactive imaging [15] mainly includes single-photon emission computed tomography (SPECT) and positron emission tomography (PET) scans. Both imaging techniques use radioisotope-labeled drugs to obtain images through the distribution of the drug in the body. SPECT is a functional imaging technique that evaluates the function of organs such as the heart, brain and tumors. PET focuses more on biochemical processes, which can be used to study brain function, tumor metabolic activities, etc. The advantage of radioactive imaging is that it can provide molecular-level images, which has good useful for studying the mechanism of disease occurrence and therapeutic effect. However, its limitation lies in the complex supply and treatment of radiopharmaceuticals, and there are certain radiation risks.

2. **The characteristics of the 2D imaging system**

2D optical imaging systems, have some remarkable features, which give them unique advantages in a variety of applications. First, 2D imaging systems enable non-contact imaging. This means that it can obtain its image information without touching the target object, thus avoiding possible physical damage and contamination. Second, 2D imaging systems usually have high resolution. Both scanning imaging optics and gaze imaging optics can capture images of a target object at a high resolution. This makes 2D imaging systems perform well in applications that require high-precision imaging. Moreover, the 2D imaging system is suitable for large-scale imaging. Additionally, the 2D imaging system also realizes the new acquisition and codec of the signal by compressed sensing, realizes the imaging with less data amount, effectively suppresses the side lobe, improves the resolution of the target in the imaging to a certain extent, and provides an effective way to effectively reduce the high resolution.

## 6.3 Resolution and Dynamic Range of 2D Imaging

1. **Resolution of the one-and 2D imaging**

(1) **Definition of Resolution**

Resolution is a key measure of the performance of a 2D imaging system. It refers to the system's ability to distinguish between details in space, spectrum, and time. In terms of spatial resolution, it reflects the system's ability to differentiate the fine details of the target object, specifically the minimum size that can be displayed. In terms of spectral resolution, it represents the system's ability to distinguish between different wavelengths of light, or the minimum spectral interval it can resolve. Temporal resolution refers to the time interval between two successive images captured by the system.

(2) **Classification of the resolution**

A. **Spatial resolution**

Spatial resolution [6–9] defines the smallest size at which an imaging system can distinguish the display and is usually expressed by full angular resolution (FWHM) or radian. The higher the spatial resolution, the clearer the details of the imaging system, which is suitable for applications requiring high-definition imaging, such as astronomical observation, biological microscopic imaging, etc.

B. **Spectral resolution**

Spectral resolution refers to the ability of imaging systems to distinguish light at different wavelengths in spectral dimensions. Systems with high spectral resolution can resolve very close spectral lines, which are crucial for research in areas such as spectral analysis, environmental monitoring, and astrophysics.

C. **Time resolution**

The temporal resolution is the time interval between two consecutive imaging sessions of the imaging system. The higher the temporal resolution, the faster the events the system can capture, which is suitable for high-speed photography, medical imaging, etc., especially when it is necessary to observe rapidly changing processes.

(3) **Methods to improve the resolution**

A. **Technological innovation**

In radar 2D imaging, the distance resolution can be effectively improved by increasing the bandwidth of the transmitted signal. This is because the distance resolution is inversely proportional to the bandwidth of the transmitted signal. Therefore, adopting the broadband signal is an important means to improve the 2D imaging resolution of the radar. In optical imaging, the use of higher-performance lenses and detectors can improve the spatial resolution of the imaging system. Moreover, the performance of imaging systems can also be further improved using advanced optical design and fabrication techniques such as micro-nano optical techniques.

B. **Image processing technology**

Super-resolution imaging is a technique to improve image resolution through software algorithms. It uses the redundant information and prior knowledge in the image to improve the detail and clarity of the images through interpolation, filtering, and reconstruction. Super-resolution imaging technology can effectively improve the resolution of 2D imaging without changing the hardware of the imaging system. In recent years, deep learning technology has made significant progress in the field of image processing. By training deep neural network models, automatic classification, recognition, enhancement and reconstruction can be realized. In 2D imaging, deep learning techniques can be used to improve the resolution and contrast of images, as well as to remove interference factors such as noise and artifacts.

C. **Multi-source data fusion**

By integrating image data from different imaging modalities (such as optical, infrared, radar, etc.), more comprehensive and accurate target information can be obtained. This multimodal fusion imaging technique can combine the advantages of various imaging modalities to improve the resolution and reliability of 2D imaging. In addition to using image data from different imaging modalities, data from different sensors (such as position, speed, acceleration, etc.) can be combined to improve the resolution of 2D imaging. These sensor data can provide information about the motion state and pose of the target, contributing to a more accurate positioning and identification of the target. It should be noted that improving the resolution of 2D imaging requires a comprehensive consideration of the hardware and software conditions of the imaging system, as well as the requirements of specific application scenarios. In practical applications, it is necessary to select appropriate methods to improve the resolution according to the specific situation.

2. **The dynamic range of 2D imaging**

(1) **Definition of the dynamic range**

Dynamic range is the ability of a digital imaging system to correctly capture from darkest to brightest scenes, often expressed numerically. In a digital camera or imaging sensor, the dynamic range defines the range of signal intensity, the ratio from the minimum to the maximum light intensity of the energy detected by the sensor. High dynamic range means that the system is better able to handle extremely bright or dark scenes without overexposure or over shading.

(2) **Factors affecting the dynamic range**

A. **The sensor performance**

The performance of the sensors is one of the key factors affecting the dynamic range. Different sensor designs (such as CCD, and CMOS) and pixel structures can have a significant impact on the dynamic range. For example, backshot (BSI) CMOS sensors typically provide a higher dynamic range than conventional CMOS sensors because they perform better in lower-light environments. Moreover, the size of the sensor also affects the dynamic range, and larger sensors are usually able to capture more light, thus providing a higher dynamic range.

B. **Optical system**

The quality of the optical system also has an important impact on the dynamic range. The optical performance of the lens, including its sharpness, contrast, and distortion, can all affect the dynamic range of the final imaging. For example, high-quality lenses can better transmit light and reduce loss, thus increasing the dynamic range. Alternatively, the design of optical filters, such as infrared and ultraviolet filters, may also affect the dynamic range of the sensor.

C. **Exposure control**

Exposure control is another key factor in adjusting dynamic range. Exposure time, aperture size, and ISO sensitivity are the three basic elements of exposure control. Longer exposure times can capture more light and improve dark detail, but may also introduce noise. A proper aperture size ensures that all areas of the scene get enough light, while an ISO sensitivity adjustment improves sensitivity with insufficient light, but improving the ISO also increases image noise.

(3) **Methods to improve the dynamic range**

A. **High dynamic range (HDR) imaging technology**

High dynamic range (HDR) imaging is a technology that captures a wider range of brightness to better reflect light conditions in the real world. This technique can save information from the darkest to the brightest light in the image, allowing the image to show more detail and reality when displayed. The core of HDR imaging technology lies in the control of the exposure time and the synthesis of images. By shooting the same scene with multiple different exposure times and combining these images, an image with a higher dynamic range.

B. **Image processing technology**

Image processing technology is another way to improve the dynamic range. This technique extends the dynamic range of images mainly by performing digital processing of them. Common image processing techniques include histogram equalization, contrast enhancement, local exposure adjustment and so on. These techniques can effectively improve the visual effect of the image, making the image more uniform in brightness, thus increasing the dynamic range.

C. **Multi-exposure and fusion**

Multi-exposure fusion is a technique that improves the dynamic range by combining multiple exposure images. This approach typically requires using a camera to take multiple photos of the same scene with different exposure times, and then fuse them with a specific algorithm. In this process, each exposure time captures the details of the different luminances in the scene, and the fusion yields an image with a higher dynamic range. The key of multi-exposure fusion technology is the correct selection of exposure time and the effective image fusion algorithm.

(4) **Application of the dynamic range**

A. **Photography**

In photography, the dynamic range is an important technical indicator. It determines the camera's ability to capture the brightest and darkest details during the shoot. High dynamic range (HDR) photography can better represent the lighting effect in the real world, especially in scenes with strong light contrast. By expanding the brightness range, HDR photography can keep both the details of the high and dark, avoiding excessive or underexposures. This technique is particularly important in landscape photography, architectural photography, and advertising photography which requires rich detail.

B. **Microscopic imaging**

Microscopic imaging [12, 16–18] plays a key role in biomedical research, materials science, and nanotechnology. The dynamic range is equally critical here because it affects the ability to observe the internal structure of cells, tissues, or materials. A wide dynamic range means that the microscope can distinguish details under extreme luminance differences, which is particularly useful for observing transparent or translucent samples. For example, while observing live cells, the light difference inside and outside the cell may be very large. The wide dynamic range ensures that the details inside and outside the cell can be captured and displayed.

C. **Remote sensing imaging**

Remote sensing imaging is a technology that collects information on the earth's surface through sensors on a satellite or an aircraft. In the field of remote sensing, the dynamic range is crucial for the correct interpretation and analysis of the reflectivity of different features. For example, in agricultural monitoring, remote sensing images with a wide dynamic range can distinguish crops with different growth conditions; in environmental monitoring, land and water bodies with different coverage types can be distinguished. In addition, for volcanoes, deserts on the earth's surface and other extreme brightness conditions, the wide dynamic range also ensures that remote sensing images can provide detailed and accurate surface information, which is of great significance for disaster warning and resource management.

3. **The relationship and trade-off between resolution and dynamic range**

High resolution usually requires more sensors or more delicate sampling techniques, which may sacrifice the dynamic range because more sampling sites absorb more light, thus reducing the sensitivity to extremely dark or extremely bright scenes. Conversely, a compromise on the sensor may be required to obtain a larger dynamic range, which may reduce the resolution.

In practical applications, such as photography, film shooting, and various monitoring devices, users often need to make a trade-off between resolution and dynamic range. For example, when shooting scenes with large contrast intensity, such as sunrise and sunset, a large dynamic range is needed to capture details from dark to bright, and some resolution may be sacrificed. In scenarios where high-definition details are required, such as print and publishing, priority should be given to resolution, and a smaller dynamic range may be accepted.

In terms of technology development, engineers are also trying to break through this trade-off, such as multi-exposure synthesis technology, and improved sensor design, to improve resolution while expanding the dynamic range. However, such technologies often bring about an increase in cost, and complexity.

Therefore, for different application scenarios, users need to select the appropriate imaging system according to their actual needs to achieve optimal imaging results. When the budget and technical conditions allow, seeking the balance between resolution and dynamic range is the key to improving the imaging quality.

## 6.4 Application of 2D Imaging in Machine Vision, Safety Inspection and Other Fields

1. **The application of one-dimensional and 2D imaging in the field of machine vision**

(1) **Overview of the machine vision system**

The machine vision system is a comprehensive technology, which converts optical signals into image signals through image acquisition hardware (such as camera, lens, light source, etc.), and transmits them to a dedicated image processing system. According to the information such as pixel brightness and color distribution, the image processing system extracts the target features and makes corresponding judgments, and then controls the field equipment according to the results to realize the detection function.

Machine vision system involves many technical fields, including image processing, mechanical engineering technology, control, electric light source lighting, optical imaging, sensors, simulation and digital video technology, computer hardware and software technology, etc. A typical machine vision application system usually includes image capture, a light source system, and an image digitization module, a digital image processing module, an intelligent judgment decision module and a mechanical control execution module.

Features of machine vision systems include increased flexibility and automation of production. In some dangerous working environments that is not suitable for manual operation or where artificial vision is difficult to meet the requirements, machine vision is often used to replace artificial vision. At the same time, during repetitive industrial production, the machine vision detection method can greatly improve the efficiency and automation of production.

The machine vision system is widely used, covering industry, agriculture, medicine, military, aerospace, meteorology, astronomy, public security, transportation, security, and scientific. For example, during the automobile manufacturing process, machine vision can be used to detect the defects on the body surface, the uniformity of the painting, and the integrity of the welding points. In addition, machine vision is also used in robot navigation and operation, commodity identification and classification, medical image analysis, and intelligent driving and unmanned driving.

With the development of technology, the machine vision system is developing towards higher accuracy, faster real-time, multi-modal fusion and intelligence. Deep learning techniques have also been applied in machine vision to enable more accurate tasks such as image recognition, object detection, and image segmentation. At the same time, with the continuous improvement of neural network models and algorithms, deep learning will play a more important role in machine vision.

(2) **Application of 2D imaging in machine vision**

A. **Automated visual detection on the production line**

2D imaging technology is widely used in the automatic production line, mainly reflected in product detection, defect detection, dimension measurement and so on. Through the high-precision 2D imaging equipment, the image information of the product can be obtained in real-time, and combined with the image processing algorithm, it can quickly and accurately judge whether the product meets the production standards. For example, in the electronic manufacturing industry, 2D imaging technology can be used to detect whether the components on the circuit board are correctly installed and whether the solder joints are qualified. In the automobile manufacturing industry, it can be used to test whether the size and shape of auto parts meet the design requirements. In addition, in food, medicine and other industries, 2D imaging technology can also be used to detect the appearance, size, and shape of products to ensure the quality of products.

B. **Robot navigation and positioning**

2D imaging technology also plays an important role in robot navigation and positioning. Through the 2D imaging equipment, the robot can obtain information about the surrounding environment, combined with the positioning algorithm, can accurately determine its location, to carry out path planning and complete the specified tasks. For example, in intelligent storage, robots need to identify items on shelves through 2D imaging technology, calculate the best pickup path, and complete the handling of goods. In the field of unmanned driving, 2D imaging technology can be used to identify obstacles and traffic signs on the road, to ensure the safe driving of vehicles.

C. **Intelligent identification and classification**

2D imaging technology is also widely used in the field of intelligent identification and classification. The image information obtained by the 2D imaging device can use the image processing algorithm to realize the recognition and classification of the objects in the image. For example, in the field of medical imaging, 2D imaging technology can be used to intelligently identify X-ray tablets and CT tablets to assist doctors in diagnosis. In the field of security, personnel can be identified and classified through 2D imaging technology to realize intelligent monitoring. In the field of logistics, 2D imaging technology can be used to intelligently classify express delivery and improve logistics efficiency.

(3) **2D imaging technology to improve machine vision performance**

A. **High-resolution imaging technology improves the detection accuracy**

High-resolution imaging technology plays a crucial role in the field of machine vision. This technique makes the image sharper by improving the imaging resolution and thus capturing the details more accurately. This is of special significance for those application scenarios that require high-precision detection, such as semiconductor manufacturing, precision machining, etc. During the semiconductor manufacturing process, high-resolution imaging technology can help machine vision systems to more accurately identify and measure the circuit patterns on the wafer, ensuring the accuracy and consistency of the circuit. In the field of precision machining, high-resolution imaging technology can help the machine vision system to more accurately measure and monitor the size and shape of the object, thus improving the accuracy and quality of machining. In addition, high-resolution imaging technology can also be applied in quality control, robot navigation, object recognition and other fields. By improving the resolution of imaging, the machine vision system can more accurately identify and process the target object, and improve the performance and efficiency of the whole system. Collectively, high-resolution imaging technology has a wide range of applications in the field of machine vision and can bring higher benefits for various application scenarios by improving detection accuracy.

B. **Dynamic range extension to enhance the brightness and contrast of the image**

Dynamic range extension is another key technology in machine vision. Practically, the brightness and contrast of images are often affected by factors such as ambient light, leading to some important details that cannot be accurately captured. Dynamic range expansion technology can effectively enhance the brightness and contrast of the image, making the dark details of the image clearer, while retaining the information of the bright part. This helps the machine vision system to obtain high-quality images under different light conditions, thus improving the stability and reliability of the whole system.

C. **Advanced technologies such as deep learning can improve the accuracy and efficiency of image recognition and classification**

With the development of advanced technologies such as deep learning, the performance of machine vision in image recognition and classification has been significantly improved. By training the deep learning model, the machine vision system can automatically learn and extract the features in the image, to realize the efficient recognition and classification of the target objects. This technology has achieved remarkable results in areas such as face recognition, object detection, and image segmentation. Simultaneously, the implementation of deep learning technology also makes the machine vision system more robust and can work normally in some complex and harsh environments.

The application of 2D imaging technology in the field of machine vision is of great significance. Through the application of advanced technologies such as high-resolution imaging technology, dynamic range expansion technology and deep learning, the performance of the machine vision system can be effectively improved, so that it can play the advantages of high efficiency and accuracy in various application scenarios.

2. **Application of 2D imaging in the field of security inspection**
(1) **Overview of the safety inspection system**

The security inspection system is an important security measure, which is widely used in various occasions, such as airports, railway stations, subway stations, large-scale activities, etc. Its main purpose is to ensure the safety of personnel and property and prevent dangerous goods and contraband from being brought into public places. The security inspection system usually consists of the security inspection machine, manual inspection, auxiliary equipment, etc. The security inspection machine is the core equipment in the security inspection system. Its main function is to scan the passing luggage or personnel to detect whether it carries contraband or dangerous goods. At present, there are two types of X-ray security check machines and 2D imaging security check machines on the market. The X-ray security inspection machine adopts the X-ray penetration principle, which can clearly show the structure inside the luggage, but the exterior details are not clear enough. The 2D imaging security check machine uses advanced 2D imaging technology, which can display the details of the luggage at the same time, greatly improving the accuracy of the security check.

(2) **Application of 2D imaging in safety inspection**
A. **Airport baggage security check**

For airport baggage security inspection, 2D imaging technology has become an important tool to improve the efficiency and accuracy of security inspection. Although traditional X-ray security technology can penetrate luggage, the identification of items depends on the density and shape of the items, often unable to clearly show details, and it is difficult to distinguish liquid, powder and other substances, which limits the accuracy of security checks to a large extent. 2D imaging technology can provide a more detailed image of the inside of the luggage, allowing security personnel to more intuitively and accurately identify the items in the luggage. 2D imaging technology usually uses X-ray computed tomography (CT) or millimeter wave imaging technology, these technologies can generate luggage internal high-resolution 3D images, security personnel cannot open the luggage, view the luggage's internal structure, effectively identify prohibited items hidden in the luggage, such as explosives, knives, guns, etc. In addition, the 2D imaging technology can also significantly improve the efficiency of security checks. Compared to traditional X-ray security equipment, 2D imaging devices can scan faster without the need to flip or reposition the luggage, greatly reducing the waiting time of passengers. For high-traffic airports, this technology is particularly important, helping to maintain the flow of security lines and improve the travel experience. In terms of security, 2D imaging also performs well. Its high-resolution images can display small and hidden threat items, which helps security personnel detect and deal with potential security risks in time, ensuring the absolute security of the airport environment. The application of 2D imaging technology in airport baggage security checks not only improves the accuracy and efficiency of security check work but also provides passengers with faster and safer services, which is an indispensable part of the modern airport security check system.

B. **Face recognition and authentication**

2D imaging technology plays an important role during face recognition. The face images obtained by 2D imaging technology have high recognition degree and can provide accurate data support for authentication. 2D imaging technology can realize the rapid capture and processing of face images, thus improving the efficiency of authentication. 2D imaging technology can effectively prevent identity fraud and forgery. Because of the high recognition of face images obtained by 2D imaging technology, even if criminals try to fake the identity of others, they are easy to be identified by the identification system, so as to improve the accuracy of security inspection. The combination of 2D imaging technology and artificial intelligence algorithms provides a real-time and efficient solution for face recognition. Artificial intelligence algorithms can quickly analyze and process the face images acquired by 2D imaging technology, to realize the real-time authentication of individuals. This technology has been widely used in security, finance, transportation, and other fields, providing a safe and reliable means of authentication for various scenarios. Although 2D imaging technology has obvious advantages during face recognition and authentication, it also faces some challenges, such as the influence of light, Angle and other factors on imaging quality. In the future, with the continuous development of technology, the application of 2D imaging technology for identity verification will be more extensive and in-depth, providing more convenience for our lives.

C. **Monitoring and security**

2D imaging technology has been widely used in monitoring and security. Its high definition and high-resolution characteristics enable the monitoring equipment to capture the detailed information in the picture in real-time, providing strong support for case detection and prevention. 2D imaging technology can realize high-definition video recording, providing a strong guarantee for post-forensics. Through this technology, the monitoring equipment can capture a clearer and more delicate picture, so that every detail in the picture can be recorded, providing strong support for the post-investigation and evidence-collection work. 2D imaging technology combined with artificial intelligence analysis can realize intelligent monitoring and improve the level of security. Through artificial intelligence analysis, the monitoring equipment can not only capture the picture, but also analyze the picture, identify the abnormal situation, and issue the alarm in time, to improve the efficiency and effect of security work. In general, the application of 2D imaging technology for monitoring and security not only improves the performance of monitoring equipment, but also improves the efficiency and effect of security work, and provides a more powerful guarantee for the safety of our life and work.

(3) **2D imaging technology enhances the effect of safety inspection**

A. **Multimodal imaging technology improves the detection capability, such as combining X-ray** [9]**, infrared and other imaging methods**

An important development direction of the application of 2D imaging technology during security inspection is the fusion of multimodal imaging technology. Traditional single imaging techniques often have some limitations, such as X-ray imaging is excellent in penetration, but cannot provide detailed information on the surface; infrared imaging can capture the thermal contour of the object, but it is difficult to analyze the internal structure of the object. By the combination of X-ray, infrared, ultraviolet, millimeter wave and other imaging technologies, the all-round and multi-level detection of the detected object can be realized, which greatly improves the accuracy and efficiency of safety inspection. For example, in airport security, in addition to using X-rays to image the internal structure of luggage, infrared imaging technology can also be used to detect temperature changes on the surface of luggage to determine whether there are hot electronic devices or other abnormal objects. At the same time, millimeter-wave imaging technology can effectively identify potentially dangerous items under clothing without violating personal privacy. Through the integrated application of these multimodal imaging technologies, security personnel can identify and handle security threats more accurately and quickly.

B. **Automate image analysis to reduce human error and missed detection rate**

The development of 2D imaging technology, combined with advanced automated image analysis systems [17, 18], can significantly reduce human error and miss rates. Through machine learning algorithms, computers can quickly identify and classify a variety of objects in the imaging data, and even small abnormalities that are invisible to the human eye can be detected in time. For example, using deep learning technology, computers can learn the characteristics of dangerous goods such as explosives and drugs from thousands of security check images, to automatically identify and label suspicious items during actual security checks. This automated analysis system not only improves the speed of security check but also reduces the missed and false detection caused by human fatigue, inattention and other factors.

C. **Big data analysis and intelligent early warning system to improve the predictability and accuracy of security inspection**

The application of big data analysis in security inspection provides a new possibility for preventing potential risks. By collecting and analyzing a large amount of security check data, a risk prediction model can be constructed to analyze the behavior patterns of passengers, to foresee and prevent potential security threats. The intelligent early warning system can monitor all kinds of information in the security check process in real-time. Once an abnormal situation is detected, the system will immediately issue an early warning and prompt the security personnel to take corresponding measures. This system not only improves the response speed but also enhances the initiative of safety checks. For example, in the security checks of large events or important meetings, the historical data analysis can predict the possible security risks among the participants, to strengthen the targeted security checks measures. In the daily security check work, the intelligent warning system can monitor the security check image in real-time, automatically identify suspicious behaviors or items, and timely notify the security personnel for processing. In general, the 2D imaging technology combined with big data analysis and intelligent early warning system not only improves the efficiency of safety inspection but also enhances the foresight and accuracy of safety inspection, providing strong technical support for the maintenance of public safety.

# References


1. Mackay C et al (2012) AOLI: adaptive optics lucky imager: diffraction limited imaging in the visible on large ground-based telescopes. Ground-based Airborne Instrum Astron IV 8446(May 2014):844621. https://doi.org/10.1117/12.925618
2. Nikzad S et al (2017) High-efficiency UV/optical/NIR detectors for large aperture telescopes and UV explorer missions: development of and field observations with delta-doped arrays. J Astron Telesc Instrum Syst 3(03):1. https://doi.org/10.1117/1.jatis.3.3.036002
3. Brown DM, Ghezzo M, Garfinkel M (1975) Transparent metal oxide electrode CID imager array. Dig Tech Pap IEEE Int Solid State Circuits Conf 1975(1):34–35. https://doi.org/10.1109/ISSCC.1975.1155356
4. Pool PJ, Burt DJ, Bell RT (2006) Electron multiplying CCDs. In: SNIC Symposium Stanford, California, pp 2–7
5. Janesick JR (2001) Scientific charge-coupled devices. SPIE press
6. Tseng E et al (2021) Neural nano-optics for high-quality thin lens imaging. Nat Commun 12(1):6493. https://doi.org/10.1038/s41467-021-26443-0
7. Li Z et al (2024) Dynamic acoustic holography: one-shot high-precision and high-information methodology. Micromachines 15(11). https://doi.org/10.3390/mi15111316
8. Kim Y-H (2007) Acoustic holography BT—Springer handbook of acoustics. In: Rossing TD (ed). Springer New York, New York, NY, pp 1077–1099
9. Vermeire B, Lee L, Parks HG, Member S (1998) The effect of copper contamination on field overlap edges and perimeter junction leakage current. IEEE Trans Semicond Manuf 11(2):232–238
10. Javaid M, Haleem A, Singh RP, Rab S, Suman R (2022) Exploring impact and features of machine vision for progressive industry 4.0 culture. Sensors Int 3:100132. https://doi.org/10.1016/j.sintl.2021.100132
11. Kalamkar S (2023) Multimodal image fusion: a systematic review. Decis Anal J 9:100327. https://doi.org/10.1016/j.dajour.2023.100327
12. Wang Y et al (2022) The development of microscopic imaging technology and its application in micro- and nanotechnology. Front Chem 10. https://doi.org/10.3389/fchem.2022.931169
13. Golding CG, Lamboo LL, Beniac DR, Booth TF (2016) The scanning electron microscope in microbiology and diagnosis of infectious disease. Sci Rep 6(1):26516. https://doi.org/10.1038/srep26516
14. Noordzij W, Glaudemans AWJM (2015) Nuclear medicine imaging techniques BT—nuclear medicine and radiologic imaging in sports injuries. In: Glaudemans AWJM, Dierckx RAJO, Gielen JLMA, (Hans) Zwerver J (eds). Springer Berlin Heidelberg, Berlin, Heidelberg, pp 25–48
15. Soldevila F, Salvador-Balaguer E, Clemente P, Tajahuerce E, Lancis J (2015) High-resolution adaptive imaging with a single photodiode. Sci Rep 5:1–9. https://doi.org/10.1038/srep14300
16. Coates CG, Denvir DJ, McHale NG, Thornbury KD, Hollywood MA (2004) Optimizing low-light microscopy with back-illuminated electron multiplying charge-coupled device: enhanced sensitivity, speed, and resolution. J Biomed Opt 9(6):1244. https://doi.org/10.1117/1.1805559
17. Vyvlečka P, Pechanec V (2023) Optical remote sensing in provisioning of ecosystem-functions analysis—review. Sensors 23(10). https://doi.org/10.3390/s23104937
18. Wong MS, Zhu X, Abbas S, Kwok CYT, Wang M (2021) Optical remote sensing BT—urban informatics. In: Shi W, Goodchild MF, Batty M, Kwan M-P, Zhang A (eds). Springer Singapore, Singapore, pp 315–344

# Chapter 7
# Application of Photoelectric Detection Technology

**Abstract** Photoelectric detection plays a crucial role in advancing remote sensing, atomic absorption spectroscopy (AAS), and mechanical quantity detection. In remote sensing, it enables precise acquisition and processing of optical, infrared, and radar signals, offering applications in Earth observation, thermal monitoring, and night reconnaissance. Challenges like enhanced sensitivity and response speed are addressed through innovations in materials, device structures, and intelligent systems. Integration with image processing and AI ensures real-time monitoring and quality assurance in production environments. The continuous advancement of high-resolution imaging technologies further enhances the accuracy and scope of photoelectric detection. These developments, driven by AI and micro-nano optical technologies, promise to expand applications in virtual reality, medical imaging, and security.

**Keywords** Photoelectric detection · Remote sensing · Atomic absorption spectroscopy · Mechanical quantity detection · High-resolution imaging · Artificial intelligence · Big data · Sustainability · Signal processing

## 7.1 Application of Photoelectric Detection in Remote Sensing

1. **Application of photoelectric detection in remote sensing**

(1) **Photoelectric detection in optical remote sensing**

A. **The application of photoelectric detectors in optical remote sensing satellites and ground stations**

Optical remote sensing [1, 2] is a remote sensing technique that obtains information by analyzing light reflected or radiated from the Earth or other celestial bodies. In optical remote sensing, the photo detector plays a crucial role. They are widely used in satellites and ground stations to detect and record optical signals in various electromagnetic spectrum segments, thus enabling the monitoring and analysis of the Earth's surface and its environment. The main function of the photoelectric detector

 
S. Du and W. Yin, *Photodetection and Image Sensing Techniques*, Advanced Topics in Science and Technology in China 2, https://doi.org/10.1007/978-981-96-5479-6_7

is to convert optical signals into electrical signals to facilitate signal transmission, processing and analysis. On satellite platforms, photo detectors are often used in combination with telescopes or other optical instruments to collect the reflected light or thermal radiation emitted from the Earth's surface. These photo detectors can be PDs, PDs, avalanche PDs, etc., which have the characteristics of high sensitivity, high resolution, fast response and so on, and can meet the needs of different application scenarios. In the ground stations, the photoelectric detector is mainly used to receive and analyze the optical signals transmitted down from the satellite. For example, in an Earth observation satellite system, a ground station photo detector can receive photos taken by a high-resolution camera on the satellite and convert them into electrical signals for further image processing and analysis.

B. **The role of photoelectric detection technology in image acquisition, transmission and processing**

In the image acquisition stage, photoelectric detection technology is the key to realizing the conversion from optical signal to electrical signal. By using a high-sensitivity photo detector, optical information in the image can be efficiently collected and recorded, resulting in obtaining high-quality image data. In addition, the photoelectric detection technology can also realize the enhancement and adjustment of images to meet the needs of different application scenarios. In the image transmission stage, the photoelectric detection technology plays the role of converting optical signals into electrical signals. By using optical fiber communication technology, electrical signals can be transmitted far away from the transmitting source, while maintaining the stability and integrity of the signal. In the image processing stage, the photoelectric detection technology can be used to realize the digital processing and analysis of the image. By using high-performance digital signal processors and algorithms, the electrical signals output by the photo detector can be processed and analyzed to identify, classify, and interpret images. Moreover, photoelectric detection technology can also be used to enhance and adjust the image to improve the quality and readability of the image.

(2) **Photoelectric detection in infrared remote sensing**

A. **Operational mechanism and characteristics of the infrared photoelectric detector**

The working principle of the infrared photoelectric detector involves the photoelectric effect, especially the interaction between the infrared radiation and the semiconductor materials. When infrared radiation hits an electrode in the detector (usually a semiconductor material), it excites electrons in the material, creating an electron–hole pair. These electron–hole pairs will separate under the electric field to form electric currents. This current can be used as a signal to record the intensity and other features of the infrared light.

The infrared photoelectric detector can obtain relevant information without direct contact with the measured object by measuring the infrared radiation emitted or reflected by the object. This makes it particularly useful when measuring high temperatures, and corrosive, or moving objects. Infrared photoelectric detectors are very sensitive to infrared radiation so even very faint infrared radiation can be detected. This makes it outstanding in applications that require high-precision measurement. Different infrared photoelectric detectors can respond to different bands of infrared radiation, which allows them to adapt to various application scenarios. The response time of infrared photoelectric detectors is usually very short and can rapidly capture and respond to changes in infrared radiation. Because of the high penetration of infrared radiation, the infrared photoelectric detector can still maintain good performance in a complex environment and is not susceptible to interference by other signals. The infrared photoelectric detector does not produce harmful radiation, which is harmless to the human body and the environment. At the same time, because it is non-contact, it also avoids the possible damage or contamination due to contact.

B. **Application of infrared remote sensing in night reconnaissance, temperature monitoring, forest fire early warning and other aspects**

The application of infrared remote sensing technology in night reconnaissance mainly depends on the high sensitivity of infrared photoelectric detectors. Through infrared remote sensing equipment mounted on reconnaissance aircraft or satellites, thermal radiation from ground targets can be detected, thus achieving night imaging and target recognition. The application of infrared remote sensing technology in temperature monitoring is realized by detecting the infrared radiation on the surface of objects. Infrared photoelectric detectors can accurately measure the temperature distribution of the surface of an object, which is widely used in meteorological observation, geological exploration and industrial detection. Forest fire early warning is one of the important applications of infrared remote sensing technology in the civil field. The infrared photoelectric detector can monitor the temperature change in the forest area in real-time. When the abnormal high-temperature area is detected, the fire warning can be issued in time, providing valuable time and information for the fire prevention department. In short, the photoelectric detection technology in infrared remote sensing plays an important role in night reconnaissance, temperature monitoring and forest fire early warning. The performance and application range of infrared remote sensing technology can be further improved through a detailed study of its working principles and characteristics.

(3) **Photoelectric detection in radar remote sensing**

A. **Photoelectric detection technology and equipment in radar remote sensing**

Radar remote sensing technology detects and measures the information of the earth's surface and its surrounding space through the radar system, and photoelectric detection technology plays a vital role in radar remote sensing. It mainly uses photo detectors to detect the echo signals of radar waves and to convert these optical signals into electrical signals to facilitate further processing and analysis. In radar remote sensing, the commonly used photoelectric detection technology and equipment mainly include PMT, PD, avalanche PD, quantum well photo detector, etc. These photo detectors have the advantages of high sensitivity, high response speed and wide band, and can effectively detect weak radar echo signals.

B. **The role of photoelectric detection in radar image generation, target recognition and information extraction**

Radar remote sensing is to obtain target information by transmitting radar waves and receiving its echo signal. Photoelectric detection technology converts the radar echo signal into an electrical signal and generates the radar image through signal processing. These images can reflect the characteristics of the position, speed, and shape of the target, and provide the basic data for the subsequent target recognition and information extraction. Target recognition in the radar images is achieved by performing feature extraction and classification of the images. Photoelectric detection technology can provide high-quality radar images, which helps to accurately extract target features and improve the accuracy and reliability of target recognition. The information extraction in radar remote sensing mainly includes classifying, detecting and tracking the targets in radar images. Photoelectric detection technology can provide high-quality radar images, help to accurately extract target information, and provide important support for environmental monitoring, resource investigation, military reconnaissance and other fields. Photoelectric detection technology plays a vital role in radar remote sensing, which not only provides the basic data for the generation of radar images but also provides high-quality image data for target recognition and information extraction. Through the continuous optimization and development of photoelectric detection technology, the performance and application effect of radar remote sensing systems can be further improved.

2. **Challenges and development trends of photoelectric detection in remote sensing technology**

(1) **Existing Challenges**

A. **Improve the sensitivity and response speed of the photoelectric detector**

In remote sensing technology, the photoelectric detector is the key component, and its sensitivity and response speed directly affect the quality and accuracy of remote sensing data. With the development of remote sensing technology, the performance requirements of photoelectric detectors are getting higher and higher. At present, there are still some limitations in the sensitivity and response speed, and how to improve its performance is an urgent problem to be solved in the development of remote sensing technology.

B. **Solve the accurate acquisition and processing of remote sensing data in complex environments**

Remote sensing technology has wide application prospects in obtaining earth surface information, but under the influence of complex environments, such as atmosphere, cloud, rain and snow and other weather conditions, the accurate acquisition and processing of remote sensing data face great challenges. How to ensure the accuracy and reliability of remote sensing data under these conditions is a problem that needs to be solved in the development of remote sensing technology.

C. **Strengthen the application of photoelectric detection technology in remote sensing data processing and analysis**

Photoelectric detection technology plays an important role in remote sensing data processing and analysis, and can effectively improve the efficiency and accuracy of data processing. However, at present, the application of photoelectric detection technology in remote sensing data processing and analysis is not extensive enough. How to strengthen its application in remote sensing data processing and analysis is a problem that needs to be considered in the development of remote sensing technology.

(2) **Development trends**

A. **Develop novel photoelectric detection materials and devices to improve the detection performance**

With the continuous progress of science and technology, the research and development of new photoelectric detection materials and devices has become an important direction of the development of remote sensing technology. For example, the use of QDs, single molecules and other nanomaterials can improve the sensitivity and selectivity of photoelectric detection. In addition, the photoelectric detection performance can be further optimized by designing new optoelectronic device structures, such as micro-nano structure, surface plasmon resonance, etc. The research of these new materials and devices will provide higher detection capability and broader application prospects for remote sensing technology.

B. **Research and development of integrated and intelligent photoelectric detection system**

To meet the application needs of remote sensing technology in complex environments, the research and development of integrated and intelligent photoelectric detection systems has become one of the development trends. The integrated photoelectric detection system can realize photoelectric detection with a wide spectral range, high sensitivity and high SNR by integrating multiple photoelectric detection units. Intelligent photoelectric detection system introduces machine learning, artificial intelligence and other technologies to achieve adaptive adjustment, automatic calibration and other functions, to improve the stability and reliability of the system. The research and development of these integrated and intelligent photoelectric detection systems will play an important role in remote sensing technology in environmental monitoring, resource exploration, national defense and security and other fields.

C. **Improve the remote sensing data processing and analysis capabilities by combining big data, artificial intelligence and other technologies**

With the wide application of remote sensing technology in various fields, the amount of remote sensing data obtained is constantly increasing, and how to efficiently process and analyze these data becomes an important challenge. Combined with big data, artificial intelligence and other technologies, the rapid processing, intelligent analysis and information extraction of remote sensing data can be realized. For example, using deep learning algorithms can realize the automatic classification and target detection of remote sensing images, and improve the degree of automation and accuracy of data processing. In addition, through the construction of a cloud computing platform for remote sensing data processing and analysis, large-scale sharing and collaborative processing of remote sensing data can be realized, and the application effect of remote sensing technology can be further improved.

The development trend of photoelectric detection in remote sensing technology is mainly reflected in the development of new photoelectric detection materials and devices, integrated and intelligent photoelectric detection system research and development, as well as the combination of big data, artificial intelligence and other technologies to improve the remote sensing data processing and analysis capabilities. These development trends will bring new opportunities and challenges to the progress and application of remote sensing technology.

## 7.2 Application of Photoelectric Detection in AAS

1. **Overview of the atomic absorption spectrum analysis**

Atomic absorption spectroscopy (AAS) [3–6], is an instrumental analysis method based on the absorption of the characteristic spectral lines of the tested elements, and then qualitative and quantitative analysis.

(1) **Definition of atomic absorption spectrum analysis**

A. **Principles of AAS**

Different elements have a certain number of electrons around their nucleus. The common and most stable orbital structures of various atoms are the ground state. When the atoms obtain a certain amount of energy, the outer electrons will be in an unstable form and become an excited state. At this point, although the atoms are unstable, they will eventually return to the ground state and emit a certain wavelength of electromagnetic wave. Atomic absorption refers to the gaseous free atoms of the same atomic emission characteristic wavelength light absorption phenomenon, AAS is based on the sample elements in the vapor ground atoms, the light source of the element of the resonance emission line absorption degree of quantitative analysis method, the wavelength area in the near ultraviolet and visible area. The schematic structure is shown in Fig. 7.1.

An atomic absorption spectrometer, also known as an absorption spectrophotometer, is mainly a single beam, or double beam, including a detector, spectroscopic system, atomizer, light source, data processing and display system. Its main function is to determine the trace amount and trace elements in the sample, which is mainly used in metallurgy, drug inspection, medical and health care, food commodity inspection, geological survey, agriculture and forestry science, environmental monitoring, water quality analysis and other fields. The advantages of this monitoring method include high accuracy, high precision, less spectral interference, good selectivity, high sensitivity, low detection limit, and can determine many kinds of elements,

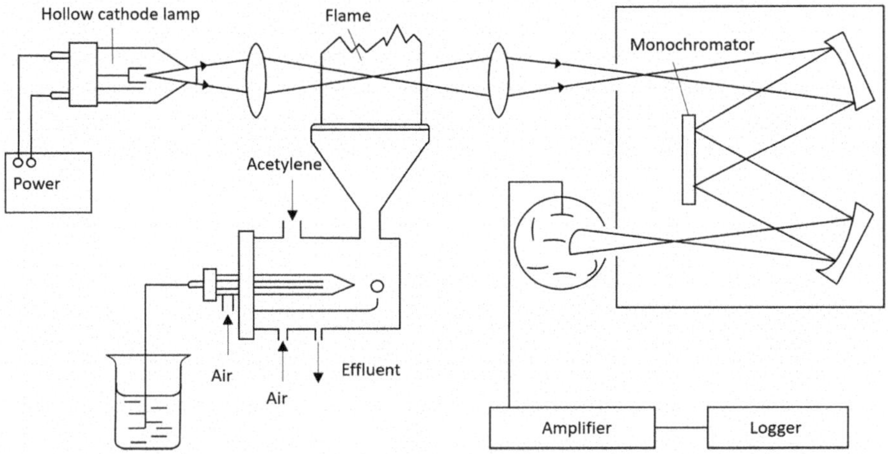


**Fig. 7.1** Schematic representation of the structure by atomic absorption spectrometry

wide application range and simple operation. But the application of this method also has certain limitations, mainly showing, for Zr, Ta, and W high-temperature element determination effect is not as good as the emission spectrum method, unable to determine a variety of elements at the same time, in the complex sample determination of the main components, when the flame in the ultraviolet zone will produce self-priming.

B. **Instruments for AAS**

Atomic absorption spectral analysis instrument principle is through the flame, graphite furnace will elements to be tested in high temperature or chemical reaction into atomic steam, by light source radiation out the characteristics of the element spectrum, through the elements of atomic steam spectral absorption, the intensity of the transmitted light is inversely proportional to the measured element concentration. In the optical path system of the instrument, the transmitted light signal is separated by a grating to separate the absorption lines of the elements to be measured from the other spectral lines. After the photoelectric converter, the optical signal is converted into the electrical signal, which is amplified and processed by the circuit system, and then analyzed and calculated by the CPU and the external computer, and finally displays the content and concentration of trace and ultra-trace metals and metal-like elements in the sample to be tested on the screen, and the printer prints the report according to the user requirements. The instrument is mainly composed of five parts: light source, atomizer, optical circuit system, circuit system and computer system.

C. **Light source**

The light source is used to produce atomic lines of the elements to be measured. It must be able to emit a narrower width than the absorption line and is a strong and stable sharp line spectrum. The commonly used light source has a hollow cathode lamp. When working, the power supply circuit of the instrument adds a voltage of 200

–500 V between the cathode and the anode of the lamp and provides different lamp operating currents according to the detection requirements of different elements. During working, the lamp was placed in the high-frequency electric field and excited by argon. As the temperature in the tube increases, the metal compound evaporates out, and further dissociates and stimulates, thus radiating out the resonance line of the metal elements.

D. **Aatomizer**

The function of the atomizer is to provide a certain amount of energy so that the elements in the sample to be tested free from the steam ground state atoms and make it enter the radiation light path of the light source, for absorption. Now the most commonly used two kinds of atomators, namely flame atomators and graphite furnaces. The flame atomator is the most commonly used atomizer, consisting of two parts: an atomizer that turns the sample solution into a highly dispersed state and a combustion head. When working, the air compressor outside the instrument provides compressed air as the combustion gas, which enters the atomizer by the pipeline and is ejected at high speed at the exit, which will cause local negative pressure, making the sample solution rise along the sample capillary under the action of atmospheric pressure, and spray into the atomization chamber together with the compressed air.

The most commonly used graphite furnace is the tube-type high-temperature graphite furnace, the graphite tube in the instrument is closed in a protective gas chamber, heating, the graphite tube inside and outside the inert gas argon. To reduce the heat radiation of the furnace body to the surrounding area, there is also cooling water outside the furnace, keeping the outside side of the atomizer below 60 °C. Graphite furnace atomizer, high atomization efficiency, less sample required, high detection sensitivity and precision.

E. **Optical separation system**

In the atomic absorption spectrum analysis, to prevent the radiation light from the absorption wavelength into the detector, a single beam separation system is used; symmetrical grating monochromator and diffraction grating as dispersion element for distribution. Through automatic wavelength selection and peak positioning by motor drive, the wavelength range of most instruments is 190–900 nm. The monochromatic light is focused by the concave mirror through the slit and illuminated on the detector.

(2) **Interference elimination and correction in atomic absorption spectrum analysis**

A. **Physical interference and elimination measures**

Physical interference refers to the sample evaporation, transfer, and atomization because the physical change of the sample will lead to a change in the absorption strength of the atoms. Physical interference belongs to non-selective interference and has the same effects on different elements. For example, the solvent vapor pressure, the solution viscosity concentration, the pressure of the atomization device, the depth of the intrusion solution, and the capillary length and diameter of the solution will all affect the injection velocity and then affect the efficiency of the atomization of the analytical elements. There are several ways to eliminate physical interference: First, The calibration curve needs to be developed when preparing and analyzing the sample standard solution with similar composition; second, Difficulties in preparing sample solutions of similar analytical composition, Standard addition method can be adopted, Improve the measurement accuracy; third, If the concentration of the analytical elements within the sample is relatively high When meeting the sensitivity requirements, The dilution solution can be eliminated and reduced by physical interference; fourthly, the ratio of absorption intensity of the elements to be measured and internal standard elements is quantified as the calibration curve, can alleviate the physical interference; fifth, When using electrothermal spectroscopy, When elements to be tested and agents are added, compounds that are difficult to volatilize, It can eliminate the physical interference during ash and drying.

B. **Chemical interference and elimination measures**

Chemical interference refers to the effect caused by chemical interactions between the sample solution or gas-phase analysis elements and coexisting substances. These interactions influence the dissociation, atomization degree, and speed of the analyzed elements, thereby reducing the absorption signal of atoms. Chemical interference is a form of selective interference that varies for different elements. The degree of interference depends on factors such as the elements being tested, the properties of interfering components, the state and position of the flame, flame temperature, flame type, coexisting components, aerosol particle size, and fog bead size.

Anionic interference occurs when anions interfere with the absorption of metal atoms. This type of interference is complex, as different anions and test elements form compounds with varying boiling and melting points, which can affect atomization. For example, phosphate ($PO_4^{3-}$) and sulfate ($SO_4^{2-}$) can inhibit the absorption of alkaline metals, cadmium, and lead. Specifically, as phosphate concentration increases, the absorbance of calcium, cadmium, and lead gradually decreases, indicating that phosphate causes negative interference with these elements.

Cationic interference refers to the effect of associated cations on the signal of the cations being measured, which can either enhance or inhibit the signal. For instance, the presence of a large amount of sodium can inhibit potassium ionization, leading to a gradual increase in potassium absorption strength.

Gas-phase interference occurs when the measured elements react with oxygen to produce oxidized vapors that are difficult to decompose, such as SiO, CeO, and $Al_2O_3$. These refractory oxides have low atomization efficiency. Using a high-temperature flame, such as one with nitrous oxide or acetylene, can improve sensitivity when measuring these elements.

Several measures can be taken to eliminate chemical interference. One approach is adding auxiliary solvents, which promote the melting and evaporation of high-melting-point substances, reducing interference and improving sensitivity. For example, when determining copper and cadmium in seawater, ammonium nitrate is used as a co-solvent to eliminate sodium chloride compounds and minimize interference between copper and cadmium. By adding ammonium nitrate ($NH_4NO_3$) to a sodium chloride solution, compounds such as $NaNO_3$ and $NH_4Cl$ are formed, which decompose at temperatures below 400 °C, thereby reducing interference.

Another method is the use of releasing agents. These agents react with interfering ions to form nonvolatile and stable compounds, thereby releasing the analyzed elements from interfering compounds. For instance, when determining calcium, lanthanum reacts with phosphate to release calcium and inhibit interference.

Chemical modifiers are also commonly used, particularly in graphite furnace spectroscopy, to suppress chemical interference. For example, when measuring cadmium (Cd), the peak absorption of Cd atoms occurs at around 600 °C, while the seawater background absorption becomes significant above 900 °C. Using a chemical modifier helps reduce background interference.

Chemical separation can also be employed to separate the analyzed elements from interfering components. This approach eliminates interference and enriches the analyte, thereby improving sensitivity. Common separation techniques include precipitation and extraction methods. Although the extraction method is widely used, it involves complex operations and requires a large number of chemical reagents, which may increase blank values, as well as result in the loss and contamination of the analyte.

Lastly, raising the flame temperature is another effective measure. Low-temperature flames often cause interference, which can be mitigated by using a high-temperature flame. Flame temperature affects the melting, evaporation, and dissociation of the sample. Most chemical interference can be eliminated by using a nitrous oxide or acetylene flame. However, higher flame temperatures may also increase ionization interference.

### C. Ionization interference and elimination measures

Ionization interference refers to the ionization initiation effect of atoms in a flame. When the analyzed element changes to free atomic ionization, resulting in a decrease in the number of ground state atoms, a decrease the absorbance value, the correction curve bends to the ordinate in the high concentration zone. In the acetylene/air flame, the ionization potential is less than 5 eV, when the ionization potential of 5.21–6.11 eV is not easy to ionize, and when the ionization potential is greater than 7.5 eV is difficult or not ionize. In the nitrous oxide/acetylene flame, the ionization degree of K, Na, Li, Rb, Cs, and other alkali metals is 98.4%, 78.9%, 63.8%, 99.1%, 99.7%, the influence of ionization should be fully considered, the uniform ionization potential is reduced, the flame temperature increases, the interference is more serious, the ionization interference methods include the following: First, control the flame temperature. When the flame temperature increases, the ionization interference increases accordingly, and the number of ionization atoms will decrease continuously with the flame temperature decrease. Controlling the flame temperature can suppress the interference, and the flame temperature can be reasonably controlled according to the ionization potential of the element. Second, add the anti-ionization agent. The standard solution and the sample add a low ionization potential ionization agent, which can provide free electrons. Usually, the lower the ionization potential, the better the interference suppression effect. Compared with the tested elements, the extinction ionization agent is lower, so it is easier to be ionized, preferentially ionized, and to suppress or eliminate the ionization of the tested elements.

At present, AAS has been widely used in water environment monitoring, soil, solid material, sediment and atmospheric environment monitoring. The content of cadmium, zinc, lead, and copper contained in water quality can be directly measured in water environment monitoring. The content of heavy metals in soil, is cadmium, lead, copper, and zinc. The content of selenium in the atmosphere can be determined by the atomic absorption spectrum method of a graphite furnace. With the continuous development of science and technology, the content of pollutants is determined by the atomic absorption spectrum method in environmental monitoring, and corresponding measures are taken to deal with it combined with the actual situation, to realize ecological protection and environmental governance.

2. **Application of photoelectric detection in AAS**

(1) **Photoelectric detection in the light source and the spectrometer**

In the light source and spectrometer, photoelectric detection plays a vital role. The principle of photoelectric detection is based on the working principle of photoelectric sensor. It detects and analyzes the characteristics of the reflection, absorption and light through the object to realize the recognition and measurement of the object.

A. **Selecting a light source to match the photoelectric detector**

In AFM [7, 8], the selection of the light source is crucial for the whole analysis process. The light source needs to have enough stability and strength to ensure that the absorption signals of the measured elements can be accurately detected. Common light sources include flame light sources, electric arc light source, and laser light sources. The flame light source is suitable for the detection of metal elements, the electric arc light source is suitable for the analysis of non-metal and transition metal elements, while the laser light source is suitable for the detection of trace elements because of its high monochrome and high intensity. The photoelectric detector is a key component in the spectrometer, and its performance directly affects the accuracy and sensitivity of spectral analysis. According to the type of light source and the requirements of analytical elements, appropriate photo detectors such as PMT, CCD and PIN PD can be selected. Among them, the PMT has the advantages of high gain and low noise, suitable for the detection of weak signals; CCD has high resolution and large dynamic range, suitable for the detection of high-intensity light signal; PIN PD is a balanced photoelectric detector with good linearly and fast response speed, and is suitable for the detection of various intensity light signals.

B. **Application and optimization of the photo detector in the spectrometer**

Photo detector is widely used in spectrometers, its main function is to detect the light emitted by the sample or the light after the sample, to obtain the composition and properties of the sample. In spectrometers, photo detectors can be divided into two categories: current sensors and photon counters. The current sensor is mainly used to detect weak light signals. Its working principle is to convert the light signal into an electrical signal through the photoelectric effect, and then amplify it through the amplification circuit, and finally obtain the intensity of the light signal through data processing. The current sensor has the advantages of high sensitivity and fast response speed, but its disadvantage is susceptible to environmental interference, and the intensity of the optical signal is limited by the receiving area of the photoelectric detector. The photon counter is mainly used to detect the optical signal of the single photon level. Its working principle is to convert the optical signal into the electrical signal through the photoelectric effect, and then count through the counting circuit, to obtain the intensity of the optical signal. Photon counters have the advantages of high sensitivity and high SNR, but their disadvantages are their high cost and require more complex data processing.

To improve the performance of the spectrometer, the photo detector is optimized. The sensitivity of the photoelectric detector can be improved by increasing the receiving area of the detector and improving the quantum efficiency of the photoelectric detector. The response speed of the photo detector can be improved by adopting a high-speed photo detector and optimizing circuit design. The noise of the photo detector can be reduced by using a low-noise amplification circuit and improving the manufacturing process of the photo detector. Data processing can be optimized by adopting an efficient data processing algorithm and optimizing the data processing process.

(2) **Atomic absorption spectrum signal detection**

A. **Characteristics of the atomic absorption signals**

Absorption analysis has extremely high sensitivity to detect very low concentrations of elements to be tested. This is mainly attributed to its unique measurement principle, namely the quantitative analysis based on the absorption of its characteristic spectral lines by the ground-state atomic vapor of the elements to be tested. Absorption spectroscopy is highly selective because each element has its unique characteristic line. This means that the content of specific elements can be accurately measured without interference from other elements. AMS has a good linear relationship over a wide range of concentrations, which makes it more flexible and convenient in practical applications.

B. **The role of the photoelectric detector in signal acquisition and conversion**

The photo detector is one of the key components in AAS analysis, and its main function is to convert optical signals into electrical signals. In AAS, when a specific wavelength of light passes through the sample, if the sample contains an element to be measured, then the element will absorb part of the light, resulting in the intensity of the transmitted light weakening. The photo detector measures this change in light intensity and converts it into an electrical signal output. Photo detectors often have the function of signal amplification to amplify weak electrical signals to a sufficient level for subsequent circuit processing and analysis. This helps to improve the sensitivity and accuracy of A spectral analysis. In the process of converting the optical signal to an electrical signal, the photoelectric detector will take a series of measures to suppress the noise interference. For example, using a high-performance PD or PMT as photosensitive elements, and using a low noise electronic circuit to process electrical signals. These measures help to improve the SNR and stability of A spectral analysis. Photo detectors play a crucial role in AAS. It can not only realize the conversion and amplification of optical signal to electrical signal but also can effectively suppress the noise interference and improve the sensitivity and accuracy of the analysis.

(3) **Data processing and analysis**

A. **Digitalization and processing of photoelectric detection signals**

Photoelectric detection plays an important role in the analysis of AAS, and its core task is to detect the atomic absorption light signal passing through the sample. In this process, the digitization and processing of the signal is crucial, because it is directly related to the accuracy and reliability of the analysis results. First, the photo detector converts the optical signal into an electrical signal, a step that usually involves photoelectric detection devices such as PMTs and PDs. These devices can convert the energy of photons into the kinetic energy of electrons, thus generating electrical signals. However, due to the influence of various noises, this electrical signal usually contains a large amount of random noise, so it requires denoising processing. Denoising processing usually includes two steps: analog denoising and digital denoising. Simulated denoising is usually achieved by designing low-noise photoelectric detection circuits, while digital denoising is achieved by digital signal processing technology. These techniques include sliding mean filtering, median filtering, and Kalman filtering, which can effectively remove random noise and improve the SNR of the signal. Moreover, to analyze the signal more accurately, the proper mathematical processing of the signal, such as integral, differential, square, etc. These treatments can enhance certain features of the signal, thus allowing us to analyze and interpret the signal more accurately.

B. **Data analysis methods and techniques**

Based on the digitization and processing of photoelectric detection signals, a series of data analysis methods and techniques are also needed to extract useful information and conduct quantitative analysis. First, a baseline correction of the signal is required to eliminate the effects of background noise and instrument response. Commonly used baseline correction methods include primary baseline correction, secondary baseline correction, etc. Next, we need peak detection and peak fitting of the signal to determine the peak position, shape, and intensity of the signal. These parameters are critical for the quantitative analysis. Commonly used peak detection and peak fitting methods include Gaussian fitting, Lorentz fitting, etc. Finally, a standard curve needs to be established based on the atomic absorption spectrum of the sample and the absorption spectrum of the standard solution. By comparing the absorption spectrum of the sample to be measured with the absorption spectrum of the standard curve, we can calculate the concentration of the target elements in the sample. In short, the application of photoelectric detection in AAS analysis involves a series of data analysis methods and technologies, such as signal digitization and processing, baseline correction, peak detection and peak fitting, and standard curve establishment. These methods and techniques are essential to improve the accuracy and reliability of the analytical results.

3. **Challenges and development of photoelectric detection in AAS analysis**

(1) **Challenges faced**

A. **Improve the sensitivity and stability of the photoelectric detector**

In AAS, the photo detector is one of the key components, and its sensitivity and stability directly affect the accuracy of the analytical results. However, current photo detectors still face some challenges in terms of sensitivity and stability. To improve the sensitivity, we need to increase the photocurrent, thus improving the detection capability of the detector. Moreover, to ensure stability, photo detector noise and drift need to be reduced. Therefore, how to improve the sensitivity while ensuring stability is an important problem to be solved in the design of the photo detector.

B. **Reduce the noise interference and improve the signal-to-noise ratio**

Noise interference is one of the main factors affecting SNR, and how to reduce noise interference and improve SNR is one of the important challenges. Noise interference mainly comes from the photoelectric device itself, power supply, environment and other factors. To reduce noise interference, the following methods can be adopted: firstly, high-quality photoelectric devices to reduce the noise of the photoelectric devices; filtering and stabilizing power supply to reduce power noise; thirdly, shielding and noise reduction techniques to reduce environmental noise.

C. **Realize fast and accurate data processing and analysis**

In the absorption analysis, the speed and accuracy of data processing and analysis directly affect the reliability of the analysis results. However, the current data processing and analysis methods still face some challenges. To realize rapid and accurate data processing and analysis, the following methods can be adopted: first, high-performance computers and algorithms to improve the speed of data processing and analysis; second, artificial intelligence and machine learning technology to improve the accuracy of data processing and analysis; third, multivariate analysis to improve the reliability of data processing and analysis.

(2) **Development trend**

A. **Development and application of novel photoelectric detection materials**

With the progress of science and technology, new photoelectric detection materials are more and more widely used in AAS. For example, nanomaterials can improve the sensitivity and selectivity of photo detectors due to their unique physical and chemical properties. In addition, semiconductor materials, such as Si and Ge are also widely used in photoelectric detection. The development and application of these new photoelectric detection materials have significantly improved the accuracy and efficiency of atomic absorption spectrum analysis.

B. **Integration and intelligence of the photoelectric detection system**

The integration and intelligence of photoelectric detection systems is one of the development trends in atomic absorption spectrum analysis. The integrated system can reduce the instrument volume and improve the measurement speed and stability. Intelligent systems can improve the accuracy and reliability of measurement by automatically adjusting parameters, self-calibration, and other functions. In addition, the intelligent system can also realize the remote transmission and processing of data through the connection with the computer network, which is convenient for users to conduct data analysis and sharing.

C. **Combine advanced data processing and analysis technology to improve analysis accuracy and efficiency**

Combining advanced data processing and analysis techniques can significantly improve the precision and efficiency of spectral analysis. For example, using wavelet transform, neural network, and other algorithms to preprocess and analyze spectral data can effectively eliminate noise and interference, and improve the clarity and accuracy of spectral signals. In addition, the use of big data cloud computing and other technologies can realize the rapid analysis and mining of a large number of spectral data, providing users with more accurate and comprehensive data interpretation.

## 7.3 Application of Photoelectric Detection in Mechanical Quantity Detection

1. **Foundation of mechanical quantity detection**

(1) **Definition and classification of mechanical quantity**

A. **Displacement, speed, acceleration, and other common mechanical quantities**

Displacement is the physical quantity describing the change in the position of an object, usually expressed by a vector, which is equal to the linear distance of the object from the initial position to the final position, and the direction from the initial position to the final position. In mechanical quantity detection, the measurement of displacement has an important significance for determining the position of the object, the trajectory of motion, etc.

Speed is a physical quantity that describes the speed and direction of an object, and is usually expressed by a vector. Its size is equal to the magnitude of the object displacement in unit time, in the same direction as the displacement direction. The measurement of speed plays an important role in understanding the state of an object's motion and controlling the motion process.

Acceleration is a physical quantity that describes the speed and direction of the object speed, which is usually expressed by a vector. Its size is equal to the amount of change in speed per unit time, with the same direction of speed change. The measurement of acceleration is of great significance in analyzing the motion process and controlling the motion process.

B. **Angle, torque, and other rotating machinery quantity**

Angle is the physical quantity that describes the degree of rotation of an object, and is usually expressed by radian or degree. In the case of mechanical quantity detection, the measurement of the angle plays an important role in understanding the rotation state of the object and controlling the rotation process.

Torque is a physical quantity describing the size of the rotational moment, usually expressed by Newton meter (N m). The measurement of torque is important for analyzing the rotation process and controlling the rotation process.

(2) **The importance of mechanical quantity detection**

A. **Quality control**

Mechanical quantity detection plays a crucial role in quality control. The accurate measurement of the size, shape, position, force, speed, and other physical quantities of the product can ensure that the quality of the product meets the predetermined standards. Any small deviation in the production process may lead to an unqualified product, which can affect the efficiency and reputation of the entire production batch and even the entire production line. Mechanical quantity testing provides real-time and accurate data to help manufacturers adjust the production process in time, improve product quality, and meet customer needs.

B. **Machine fault diagnosis**

Mechanical quantity detection is also an important tool for machine fault diagnosis. Through the monitoring of various physical quantities in the operation of the machine, we can find the abnormal conditions in the machine operation, such as vibration, temperature change, pressure fluctuation, etc., which may be the precursors of the machine failure. Through timely detection and analysis, the failure of the machine can be predicted, to take measures to avoid sudden damage of the machine, reduce downtime, and prolong the service life of the machine.

C. **Automatic production control**

In automated production, mechanical quantity detection is the basis for realizing accurate control. Automated production lines require accurate physical quantity data to adjust the production speed, strength, location, etc., to achieve accurate manufacturing of products. At the same time, mechanical quantity testing can also be used to monitor the production process, to ensure the stability of the production process and the quality of the product. In automatic production, mechanical quantity detection is not only the guarantee of production efficiency and product quality but also the basis of realizing intelligent production.

2. **The application of photoelectric detection in mechanical quantity detection**

(1) **Displacement and velocity measurement**

A. **Application of photoelectric encoder in displacement and velocity measurement**

The photoelectric encoder is a sensor that uses the photoelectric effect for measurement, which is widely used in the measurement of displacement and velocity. It converts mechanical displacement into electrical signals through the photoelectric effect and then obtains the displacement and speed information. In displacement measurement, the photoelectric encoder can accurately measure the position of the object, and its measurement accuracy can reach the micron level. In the velocity measurement, the photoelectric encoder can measure the linear speed of the object, and the angular speed and rotational speed of the object can be obtained through the calculation. The advantages of photoelectric encoders are their simple structure, high measurement accuracy, fast reaction speed, and strong anti-interference ability, so they are widely used in displacement and speed measurement.

B. **Application of laser interferometer in precision displacement measurement**

A laser interferometer is an instrument using the principle of laser interference. Its measurement accuracy can reach the nanometer level, so it is widely used in precision displacement measurement. The working principle of the laser interferometer is to obtain the displacement information of the object through the interference of the laser beam. When the laser beam passes through an object, the phase will change due to the displacement of the object, and the displacement information of the object can be obtained by measuring the phase change. The advantages of a laser interferometer lie in its high measurement precision, strong anti-interference ability, and large measurement range, so it is widely used in precision displacement measurement.

C. **Application of image processing technology in displacement and speed detection**

The image processing technology is a kind of image processing and analysis using a computer, which has been widely used in displacement and speed detection. In displacement detection, the position change of an object can be measured by image processing technology to obtain the displacement information of the object. In speed detection, the motion speed of an object can be measured by image processing technology to obtain the speed information of the object. The advantages of image processing technology are that it can be monitored in real-time, with high measurement accuracy and strong anti-interference ability, so it is widely used in displacement and speed detection.

(2) **Angle and rotation measurement**

A. **Application of photoelectric rotary encoder in angle measurement**

Photoelectric rotary encoder is a device that transforms rotational motion into an electrical signal, widely used in various situations where rotation angle requires accurate measurement. It is mainly composed of the optical emitter, optical receiver, and signal processing circuit. When the grating of the encoder circles, the light emitted by the light emitter is illuminated on the light receiver through the slit on the grating disk, and the light receiver generates pulse signals according to the occlusion and penetration of the light. After these pulse signals are processed by the signal processing circuit, a digital signal with a rotation angle can be obtained. It is widely used in mechanical quantity detection. For example, in CNC machine tools, robots, autonomous vehicles, and other fields, the use of photoelectric rotary encoders is needed to accurately measure the rotation angle to achieve precise control and adjustment.

B. **Application of laser gyroscope in rotational velocity and angle measurement**

A laser gyroscope [9–12] is an angular velocity sensor based on the laser principle that can be used to measure the rotation velocity and angle. It is mainly composed of lasers, optical systems, photo detectors, etc. When the laser emitted by the laser is illuminated on the rotating gyroscope through the optical system, the laser has a Doppler effect, causing a change in the light intensity received by the photo detector. Analysis of this change in light intensity yields information about the rotation velocity and angle. Because of its characteristics of high measurement accuracy, fast response speed, and strong anti-interference ability, the laser gyroscope has been widely used in aerospace, military, automobile, and other fields. For example, in an aircraft's inertial navigation system, it is necessary to use a laser gyroscope to accurately measure the rotation speed and angle of the aircraft to achieve precise navigation and control.

(3) **Vibration and acceleration measurement**

A. **Application of photoelectric sensors in vibration measurement**

Photoelectric sensors are widely used in vibration measurement. Its main function is to convert the vibration signal into the electrical signal through the photoelectric effect, to realize the measurement of the vibration speed, displacement, and other parameters. In practical applications, photoelectric sensors are usually used with optical components such as light sources and gratings to form photoelectric vibration sensors. The working principle of a photoelectric vibration sensor is: that when vibration occurs, the grating in the sensor will be driven by the vibrating object, thus changing the relative position between the grating and the light source, resulting in the beam being modulated. The modulated beam is converted into electrical signals through a photo detector, which is amplified and processed to obtain a signal related to the vibration parameters. The advantages of photoelectric sensors are their high sensitivity, high accuracy, and good linear characteristics. In addition, it also has the characteristics of non-contact measurement and is not susceptible to environmental interference, so it is widely used in vibration measurement.

B. **Application of photoelectric accelerometer in acceleration measurement**

A photoelectric accelerometer is an accelerometer measuring instrument based on the photoelectric effect. It is mainly composed of photo detectors, optical systems, and signal-processing circuits. When the acceleration of the object changes, the light intensity received by the photo detector will change, thus measuring the acceleration. Photoelectric accelerometers have a wide range of applications, including aerospace, automotive engineering, machinery manufacturing, seismic monitoring, and other fields. Its main advantages are high sensitivity, high precision, and fast response characteristics. In addition, the photoelectric accelerometer also has the advantages of being small in size, lightweight, and easy to integrate, which provides great convenience for acceleration measurement. In practical applications, photoelectric accelerometers usually use laser interferometry and beam offset optical methods to measure acceleration. These methods have high measurement accuracy and can achieve real-time, online measurement. By combining the photoelectric accelerometer with the computer and the corresponding software, the real-time display, analysis, and processing of the acceleration signal can be realized, providing powerful technical support for engineering applications.

(4) **Application of image processing in mechanical quantity detection**

A. **Application of machine vision system in part size and shape detection**

Machine vision system plays a crucial role in part size and shape detection. Through image processing technology, the system can accurately measure the size and shape of the parts, thus ensuring that it meets the predetermined standards. First, the machine vision system captures an image of the part and converts it into a digital signal. Then, through the image processing algorithm, the image can be processed to extract the size and shape information of the part. For example, you can use an edge detection algorithm to find the outline of a part and calculate its dimensions. A morphological algorithm can also be used to fill the internal gaps of the part and calculate its shape. In addition, the machine vision system can also use 3D reconstruction technology to model the parts in 3D, to more accurately measure their size and shape. This method can photograph all angles of the part and combine it into a 3D model to provide a more comprehensive understanding of the shape of the part. Overall, the application of machine vision systems in part size and shape detection can not only improve the accuracy of detection but also improve the speed and efficiency of detection.

B. **Application of high-speed camera technology in motion analysis**

High-speed camera technology [13] is a technology that can capture motion images at extremely high speed and is widely used in motion analysis. Through high-speed camera technology, images of moving objects can be captured and analyzed in detail. For example, parameters such as velocity, acceleration, and displacement can be calculated to give a precise description of its motion state. In addition, the image processing technology can also be used to track moving objects and calculate their motion trajectory. The application of high-speed camera technology in motion analysis can be used not only in sports but also in manufacturing and scientific research. For example, in the manufacturing industry, the movement of the machine can be monitored by high-speed camera technology to optimize its performance and improve its efficiency. In scientific research, the motion phenomenon in nature can be observed and analyzed by high-speed camera technology to understand its mechanism. In general, the application of high-speed camera technology in motion analysis can provide high-precision motion data, and thus provide strong support for motion analysis and research.

3. **Challenge and development of photoelectric detection in mechanical quantity detection**

(1) **Existing Challenges**

A. **Environmental light interference and noise problems**

Photoelectric detection technology is widely used in mechanical quantity detection, but environmental light interference and noise problems have been one of the main challenges. Changes in the ambient light can cause fluctuations in the optical signal, which can affect the accuracy of the detection results. Especially in the low light environment, the ambient light interference will be more obvious, resulting in the SNR of the detected signal. In addition, the noise problem is also a key factor affecting the accuracy of the photoelectric detection, and these noise sources can affect the measurement sensitivity and stability of the photoelectric detection system.

B. **Technical problems of high-precision measurement**

High-precision measurement is an important application of photoelectric detection in mechanical quantity detection. Due to the limitations of the photoelectric detection system itself, such as the responsiveness, linearity degree, and photoelectric conversion efficiency of the photoelectric detector, it will affect the high-precision measurement. In addition, the stability of the optical system will also affect the high-precision measurement. Therefore, how to improve the overall performance of the photoelectric detection system to achieve high-precision measurement is an important problem to be solved in the development of photoelectric detection technology.

C. **Requirements for real-time performance and dynamic measurement capability**

In practical application, photoelectric detection technology not only needs to meet the requirements of high-precision measurement but also needs to have real-time and dynamic measurement capability. However, this requirement is a big challenge for photoelectric detection technology. Due to the limited response speed of photoelectric detection systems, it is difficult to achieve rapid dynamic measurements. At the same time, the real-time performance requires the photoelectric detection system to complete the measurement and give the results in a short time, which puts forward high requirements for signal processing and data transmission speed. Therefore, how to improve the real-time performance and dynamic measurement capability of photoelectric detection technology to meet the needs of practical applications is another important challenge to overcome in the development of photoelectric detection technology.

(2) **Development trend**

A. **Development and application of new photoelectric detection devices**

With the continuous progress of science and technology, the research and development of new photoelectric detection devices has become an important trend of photoelectric detection technology in mechanical quantity detection. At present, people are developing new photoelectric detection devices with higher sensitivity, higher resolution, and more rapid response. For example, photoelectric detection devices based on nanotechnology, with a larger specific surface area, can achieve higher detection sensitivity; photoelectric detection devices based on new materials, with faster response speed, can achieve real-time monitoring. The application of these new photoelectric detection devices can not only improve the accuracy and real-time performance of mechanical quantity detection but also expand the application of photoelectric detection technology in more fields.

B. **Integration and intelligence of the photoelectric detection system**

The integration and intelligence of photoelectric detection systems is another important trend of photoelectric detection technology in mechanical quantity detection. With the continuous development of microelectronics technology, the photoelectric detection system is gradually changing from an independent module to an integrated system. This integrated photoelectric detection system not only has higher performance, but also has a smaller volume, lower power consumption, and easier to achieve multi-functional integration. At the same time, intelligence is also an important development direction of photoelectric detection systems. Through the introduction of artificial intelligence technology, the functions of adaptive regulation, adaptive learning, and adaptive optimization of the photoelectric detection system can be realized, to improve the intelligence level of the photoelectric detection system.

C. **Data processing and analysis methods combining artificial intelligence and big data technology**

In the photoelectric detection technology, data processing and analysis is a very important link. With the wide application of photoelectric detection technology in mechanical quantity detection, the amount of data generated is also increasing. Therefore, how to effectively process and analyze these data has become an important challenge in the development of photoelectric detection technology. Combining artificial intelligence and big data technology, rapid, accurate, and efficient processing and analysis of photoelectric detection data can be realized. For example, intelligent identification and classification of photoelectric detection data, and real-time storage, query, and analysis of photoelectric detection data. The application of these data processing and analysis methods can not only improve the performance of photoelectric detection technology but also expand the application field of photoelectric detection technology.

## 7.4 Frontier and Development Trend of Optoelectronic Imaging Technology

1. **The frontier dynamics of optoelectronic imaging technology**

The frontier dynamics of photoelectron imaging technology mainly cover new photo detectors, high-resolution imaging technology, 3D imaging technology, and computational imaging technology. Here is a brief overview of these areas:

(1) **Novel photo detectors**

Novel photo detectors are an important part of optoelectronic imaging technologies, and their development is crucial to improving the performance of imaging systems. In recent years, with continuous advances in materials science and nanotechnology, remarkable progress has been made in the study of new photo detectors. For example, new devices such as QD photo detectors and graphene-based photo detectors show higher sensitivity, wider response wavelength range, and lower noise levels. These novel photo detectors are expected to be widely used in future imaging systems. QD photo detectors [14–28], exploit the unique properties of Qs to achieve detection with high sensitivity and wide response wavelength range. QDs are semiconductor nanoparticles with sizes smaller than 10 nm. Due to the small size of the QD, the electrons are affected by the quantum limit domain effect, which is discretized. This discretized energy level structure gives the QD photo detector high sensitivity and a wide response wavelength range when detecting optical signals. Graphene-based photo detectors take advantage of the unique 2D material properties of graphene to achieve high sensitivity and low noise detection. Graphene is a 2D material composed of a single layer of carbon atoms, with extremely high electron mobility and excellent electrical conductivity. The graphene-based photo detectors can enable high sensitivity and low noise detection of optical signals by combining graphene with other materials. In addition, there are other new photo detectors, such as perovskite photo detectors, metal–organic frame photo detectors, etc. These new photo detectors all have different characteristics and advantages and are expected to be widely used in different application scenarios. The development of new photo detectors is important for improving the performance of optoelectronic imaging systems. With the continuous advances in materials science and nanotechnology, we can expect more research and application of new high-performance photo detectors to bring more possibilities for the development of optoelectronic imaging technologies.

(2) **High-resolution imaging technology**

High-resolution imaging technology is an important research direction in photoelectron imaging techniques. With the continuous development of science and technology, people's requirements for image resolution are becoming higher and higher, so high-resolution imaging technology has been widely used in various fields. Super-resolution imaging is a technique that processes low-resolution images through software algorithms to improve the detail and clarity of the images. This technology can improve the resolution of images without changing the image sensor. There are two kinds of super-resolution imaging technologies: one is super-resolution imaging technology based on image processing algorithms, and the other is based on machine learning. Both of these technologies have good application prospects. Another technique to achieve high-resolution imaging is super lenses based on micro-nano optical technology. Hyper lens is a novel imaging device that has a smaller size and higher imaging quality than conventional lenses. The realization of super lens mainly relies on micro-nano processing technology, by processing the material on a small scale, to realize the focus and control of light. The advent of super lenses offers new possibilities for high-resolution imaging. In addition, there are some other high-resolution imaging technologies, such as holographic imaging technology, fluorescence microscopy imaging technology, etc. Holographic imaging technology can realize 3D imaging of objects by recording the phase information of optical waves. Fluorescent microscopic imaging enables high-resolution imaging of biological samples by using fluorescent probes. Overall, high-resolution imaging technology plays an important role in optoelectronic imaging techniques. By continuously exploring new imaging techniques and methods, we can achieve higher image resolution, and thus meet people's demand for image quality.

(3) **3D imaging technology**

3D imaging technology is one of the hot spots of optoelectronic imaging technology in recent years, with a wide range of applications, including games, film and television, virtual reality, and other industries [29]. At present, 3D imaging technology mainly includes 3D imaging based on structured light, 3D imaging based on time-of-flight measurement, and 3D imaging based on computational imaging. The 3D imaging technology based on structured light emits the structured light to the measured object and uses the structural light reflection information on the surface of the measured object, collects through specific sensors, and then processes with the computer algorithm, to obtain the 3D form of the measured object. The advantage of this method is that high precision 3D data is suitable for high precision 3D modeling and measurement. However, its disadvantage is the need for special hardware equipment, high cost, and sensitivity to ambient light. The 3D imaging technology based on time-of-flight measurement is to calculate the 3D position of the measured object by measuring the time required for the light wave to emit from the light source to the surface of the measured object and reflect it to the sensor. The advantages of this method are that it requires no special hardware equipment, has low cost, and is insensitive to ambient light. But its disadvantage is that the accuracy is low, suitable for the accuracy requirement is not high field. 3D imaging technology based on computational imaging uses a computer algorithm to process the collected 3D image of the measured object. The advantages of this method are simple equipment, low cost, and insensitive to ambient light. However, its disadvantage is that the accuracy is low, and the computational complexity is high, which is suitable for the accuracy requirement is not high. In the future, with the continuous progress of technology and the continuous expansion of application scenarios, 3D imaging technology is expected to be applied in more fields. For example, in virtual reality, 3D imaging can provide a more realistic, immersive experience; in film production, 3D imaging can provide more realistic visual effects; in game development, 3D imaging can provide a more realistic, stereoscopic gaming experience. At the same time, with the continuous development of 3D imaging technology, its accuracy, speed, and stability performance will be further improved, to meet the needs of more application scenarios.

(4) **Computational imaging technology**

Computational imaging is a technique that applies computer vision and image processing techniques to imaging systems. It can optimize and control the imaging process through the software algorithm and improve the performance and stability of the imaging system. In recent years, with the continuous development of artificial intelligence technologies such as deep learning, computational imaging technology has also made new progress. Super-resolution imaging technique based on deep learning is an important computational imaging technique. It can obtain high-resolution images by performing deep learning processing of low-resolution images. The technology has been widely used in mobile phone cameras, satellite imaging, and other fields. Denoising and deblurring are also important components of computational imaging techniques. Image quality and clarity can be improved by denoising and deblurring the images. These technologies have important application value in medical imaging, security monitoring, and other fields. In addition, the computational imaging technology can also be combined with new photo detectors and high-resolution imaging technologies to achieve more advanced and complex imaging systems. For example, combining computational imaging systems with quantum detectors, and combining computational imaging technology with micro-nano optical technology, ultra-lightweight imaging systems. The development of computational imaging technology has brought new opportunities and challenges to imaging systems. In the future, with the further development of artificial intelligence technology, computational imaging technology will be more perfect, providing more efficient and high-quality imaging solutions for various application scenarios.

2. **The development trend of optoelectronic imaging technology**

(1) **Integration and miniaturization**

Photoelectric imaging technology has shown an obvious development trend of integration and miniaturization in recent years. With the continuous progress of science and technology, optoelectronic imaging devices are gradually developing from a single-function to a multi-function integration orientation. In this process, the performance of optoelectronic imaging technology has been significantly improved, and it also provides more possibilities for the application of optoelectronic imaging devices in various fields. In terms of integration, optoelectronic imaging technology is realizing an integrated design by integrating multiple functional units into one chip. This integrated design can not only effectively reduce the volume and weight of the device, and reduce the power consumption, but also improve the performance and reliability of optoelectronic imaging devices. For example, integrating photo detectors, signal processing circuits, memory, and so on. On a single chip can achieve high-speed, high-sensitive, and low-noise imaging performance. In addition, the integrated design can also easily achieve compatibility and integration with other systems, and further expand the application field of optoelectronic imaging technology. In terms of miniaturization, optoelectronic imaging technology is realizing the miniaturization of imaging devices through microelectronics technology and micro-electromechanical system (MEMS) technology. Miniaturized optoelectronic imaging devices have the advantages of small size, lightweight, low power consumption, and low cost, and are very suitable for portable, wearable, UAV, and other micro-nano devices. At the same time, miniaturized optoelectronic imaging devices can also achieve high-resolution and high-speed imaging, providing high-performance imaging solutions for miniature devices. For example, miniature optoelectronic imaging sensors can be applied to mobile phone cameras, visual systems for driverless vehicles, medical scopes, and other endoscopes.

(2) **Intelligence and automation**

With the development of artificial intelligence and big data technology, optoelectronic imaging technology is developing in the direction of intelligence and automation. Intelligent and automated optoelectronic imaging technology can improve the autonomous adaptation ability of the imaging system, realize the rapid identification and analysis of complex scenes and goals, and provide more efficient and accurate imaging solutions for various application scenarios. In terms of intelligence, optoelectronic imaging technology can optimize the real-time image quality in the imaging process by introducing an artificial intelligence algorithm. For example, the noise reduction and sharpening processing through the deep learning algorithm can effectively improve the sharpness and contrast of the image. In addition, the intelligent optoelectronic imaging technology can also realize the automatic focus, automatic exposure, and other functions for complex scenes, and improve the autonomous adaptability of the imaging system. In terms of automation, optoelectronic imaging technology can realize the automatic adjustment and optimization of the imaging process by introducing automatic control technology. For example, by automatically adjusting the parameters of exposure time and aperture size, the imaging system can achieve optimal imaging results in different lighting conditions and scenarios. In addition, the automated optoelectronic imaging technology can also realize the remote control and monitoring of the imaging equipment, and improve the stability and reliability of the imaging system.

(3) **Interdisciplinary integration**

The development of optoelectronic imaging technology is no longer limited to the traditional fields of electronics and optics but has begun to deeply cross and integrate with other disciplines, to realize technological innovation and expansion of application fields. Among them, the integration trend of nanotechnology and biotechnology is particularly obvious. For nanotechnology, optoelectronic imaging technology can use the high resolution and high sensitivity of nanomaterials to achieve a deeper observation and understanding of nanomaterials. For example, optoelectronic imaging devices prepared from nanomaterials can realize the imaging of individual molecules or cells, thus providing powerful tools for research in biology, chemistry, and other fields. In biotechnology, optoelectronic imaging techniques can be used to image biological samples at high resolution to achieve a precise understanding of biological structure and function. For example, using optoelectronic imaging technology can achieve high-resolution imaging of biological cells such as neurons and muscle cells, thus providing important research means for neuroscience, muscle research, and other fields. Interdisciplinary integration has played an important role in promoting the innovation and application field expansion of optoelectronic imaging technology. First, interdisciplinary integration provides new research perspectives and ideas for optoelectronic imaging technology, thus promoting technological innovation. For example, through the integration with nanotechnology, optoelectronic imaging technology has achieved a more profound observation of the microscopic world, thus driving the development of optoelectronic imaging technology. Secondly, interdisciplinary integration offers new possibilities for the application fields of optoelectronic imaging technologies. For example, through the integration with biotechnology, photoelectron imaging technology can be used to image biological samples at high resolution, thus providing important tools for research in biology, medicine, and other fields. In general, the trend of interdisciplinary integration not only promotes the innovation of optoelectronic imaging technology but also provides new possibilities for the application field of optoelectronic imaging technology, thus making an important contribution to the development of human society.

(4) **Environment-friendly and sustainable development**

With the continuous progress of science and technology, optoelectronic imaging technology plays an increasingly prominent role in environmental protection and sustainable developmen. First of all, optoelectronic imaging technology is more and more inclined to use environmentally friendly, renewable materials, such as organic light-emitting diode (OLED) screens [30], which can be naturally decomposed under certain conditions to reduce environmental pollution. Secondly, the design of optoelectronic imaging equipment is more attention to thin, and small, to reduce the use of materials and reduce energy consumption. In addition, the optoelectronic imaging technology continuously optimizes the display effect, improving the display efficiency, and thus reducing energy consumption. First of all, in terms of reducing energy consumption, optoelectronic imaging technology greatly reduces the energy consumption of the equipment by improving the display efficiency and optimizing the circuit design. For example, displays with organic OLED technology can save more energy than traditional LCD screens due to their self-emitting properties. In addition, with the development of optoelectronic imaging technology, the power consumption of the device is getting lower and lower, which makes it more widely used in mobile devices, further reducing the overall energy consumption. Second, the environmental characteristics of optoelectronic imaging technology play an important role in reducing pollution. First of all, optoelectronic imaging equipment in the production process, more and more attention to the use of environmentally friendly materials, reduces the use of harmful substances, and reduces the pollution in the production process. Secondly, the waste disposal of optoelectronic imaging equipment has also received more and more attention. For example, discarded displays can be recycled, and valuable materials can be reused, reducing resource waste and environmental pollution. The development trend of optoelectronic imaging technology in environmentally friendly and sustainable development has a positive role in reducing energy consumption and reducing pollution. With the increasing progress of technology, it is reasonable to believe that optoelectronic imaging technology will play a greater role in environmental protection and sustainable development.

## References


1. Vyvlečka P, Pechanec V (2023) Optical remote sensing in provisioning of ecosystem-functions analysis—review. Sensors 23(10). https://doi.org/10.3390/s23104937
2. Wong MS, Zhu X, Abbas S, Kwok CYT, Wang M (2021) Optical remote sensing BT—urban informatics. In: Shi W, Goodchild MF, Batty M, Kwan M-P, Zhang A (eds) Springer Singapore, Singapore, pp 315–344
3. Wang Y et al (2022) The development of microscopic imaging technology and its application in micro- and nanotechnology. Front Chem 10. https://doi.org/10.3389/fchem.2022.931169
4. Kubota R, Tanaka W, Hamachi I (2021) Microscopic imaging techniques for molecular assemblies: electron, atomic force, and confocal microscopies. Chem Rev 121(22):14281–14347. https://doi.org/10.1021/acs.chemrev.0c01334
5. Bullen A (2008) Microscopic imaging techniques for drug discovery. Nat Rev Drug Discov 7(1):54–67. https://doi.org/10.1038/nrd2446
6. Liu X et al (2023) High-performance broadband graphene/silicon/graphene photodetectors: from x-ray to near-infrared. Appl Phys Lett 122(7). https://doi.org/10.1063/5.0132028
7. Streed EW, Jechow A, Norton BG, Kielpinski D (2012) Absorption imaging of a single atom. Nat Commun 3(1):933. https://doi.org/10.1038/ncomms1944
8. Li H, Wu J, Yin Z, Zhang H (2014) Preparation and applications of mechanically exfoliated single-layer and multilayer MoS2 and WSe2 nanosheets. Acc Chem Res 47(4):1067–1075. https://doi.org/10.1021/ar4002312
9. Huang Z, Oh J, Banerjee SK, Campbell JC (2007) Effectiveness of SiGe buffer layers in reducing dark currents of Ge-on-Si photodetectors. IEEE J Quantum Electron 43(3):238–242. https://doi.org/10.1109/JQE.2006.890395
10. Filatov YV, Pavlov PA, Velikoseltsev AA, Ulrich Schreiber K (2020) Precision angle measurement systems on the basis of ring laser gyro. Sensors (Switzerland) 20(23):1–14. https://doi.org/10.3390/s20236930
11. Hurst RB, Mayerbacher M, Gebauer A, Schreiber KU, Wells J-PR (2017) High-accuracy absolute rotation rate measurements with a large ring laser gyro: establishing the scale factor. Appl Opt 56(4):1124. https://doi.org/10.1364/ao.56.001124
12. Pisani M et al (2018) G-LAS: a ring laser gyroscope for high accuracy angle measurements. J Phys Conf Ser 1065(3). https://doi.org/10.1088/1742-6596/1065/3/032009
13. Zou SJ, Shen Y, Xie FM, De Chen J, Li YQ, Tang JX (2020) Recent advances in organic light-emitting diodes: toward smart lighting and displays. Mater Chem Front 4(3):788–820. https://doi.org/10.1039/c9qm00716d
14. Muchahary D, Maity S (2022) A qualitative analysis of the impact of P composition in GaAs1−xPx capped InAs/GaAs quantum dot hetero-structure. Micro Nanostruct 164. https://doi.org/10.1016/j.spmi.2022.107154
15. Wang Z, Gu Y, Li X, Liu Y, Liu F, Wu W (2023) Recent progress of quantum dot infrared photodetectors. Adv Opt Mater 11(19):2300970. https://doi.org/10.1002/adom.202300970
16. Haider G et al (2016) Electrical-polarization-induced ultrahigh responsivity photodetectors based on graphene and graphene quantum dots. Adv Funct Mater 26(4):620–628. https://doi.org/10.1002/adfm.201503691

17. Xu Y et al (2017) Graphene coupled with silicon quantum dots for high-performance silicon Schottky photodetectors. In: 2016 13th IEEE international conference solid-state and integrated circuit technology ICSICT 2016—proceedings, pp 820–823. https://doi.org/10.1109/ICSICT.2016.7999051
18. Gan X et al (2013) Chip-integrated ultrafast graphene photodetector with high responsivity. Nat Photonics 7(11):883–887. https://doi.org/10.1038/nphoton.2013.253
19. Lin CH, Liu CW (2010) Metal-insulator-semiconductor photodetectors. Sensors 10(10):8797–8826. https://doi.org/10.3390/s101008797
20. Ding N et al (2022) A novel approach for designing efficient broadband photodetectors expanding from deep ultraviolet to near infrared. Light Sci Appl 11(1). https://doi.org/10.1038/s41377-022-00777-w
21. Guo W, Xu S, Wu Z, Wang N, Loy MMT, Du S (2013) Oxygen-assisted charge transfer between ZnO quantum dots and graphene. Small 9(18):3031–3036. https://doi.org/10.1002/smll.201202855
22. Nikitskiy I et al (2016) Integrating an electrically active colloidal quantum dot photodiode with a graphene phototransistor. Nat Commun 7(1):11954. https://doi.org/10.1038/ncomms11954
23. Sun Z, Liu Z, Li J, Tai GA, Lau SP, Yan F (2012) Infrared photodetectors based on CVD-grown graphene and PbS quantum dots with ultrahigh responsivity. Adv Mater 24(43):5878–5883. https://doi.org/10.1002/adma.201202220
24. Konstantatos G et al (2012) Hybrid graphene-quantum dot phototransistors with ultrahigh gain. Nat Nanotechnol 7(6):363–368. https://doi.org/10.1038/nnano.2012.60
25. Ni Z et al (2017) Plasmonic silicon quantum dots enabled high-sensitivity ultrabroadband photodetection of graphene-based hybrid phototransistors. ACS Nano 11(10):9854–9862. https://doi.org/10.1021/acsnano.7b03569
26. Du S et al (2017) Graphene/silicon-quantum-dots/Si Schottky-PN cascade heterojunction for short-wavelength infrared photodetection. In: IEEE international electron devices meeting (IEDM), no 2, pp 207–210
27. Angizi S et al (2022) A comprehensive review on planar boron nitride nanomaterials: from 2D nanosheets towards 0D quantum dots. Prog Mater Sci 124(April 2020):100884. https://doi.org/10.1016/j.pmatsci.2021.100884
28. Gao Z (2023) Application of 3D virtual reality technology in film and television production under internet mode BT—cyber security intelligence and analytics, pp 341–349
29. Bao Y (2022) Application of virtual reality technology in film and television animation based on artificial intelligence background. Sci Program 2022(1):2604408. https://doi.org/10.1155/2022/2604408
30. Song Q, Wook YS (2020) Exploration of the application of virtual reality and internet of things in film and television production mode. Appl Sci 10(10). https://doi.org/10.3390/app10103450

The manufacturer's authorised representative in the EU is Springer
Nature Customer Service Centre GmbH, Europaplatz 3, 69115 Heidelberg,
Germany. If you have any concerns regarding our products, please
contact ProductSafety@springernature.com

Printed and bound by CPI Group (UK) Ltd, Croydon, CR0 4YY
01/10/2026
02225935-0004